VI International Congress la ValSe-Food

VI International Congress la ValSe-Food

Volume Editors

Nancy Chasquibol
Norma Sammán
Pedro Maldonado
Laura Mereles
Ma. Carolina Zúñiga-López
Ritva Repo-Carrasco-Valencia

MDPI

Basel • Beijing • Wuhan • Barcelona • Belgrade • Novi Sad • Cluj • Manchester

Volume Editors

Nancy Chasquibol
Universidad de Lima
Lima
Peru

Norma Sammán
Universidad Nacional de Jujuy
San Salvador de Jujuy
Argentina

Pedro Maldonado
Escuela Politécnica Nacional
Quito
Ecuador

Laura Mereles
Universidad Nacional de Asunción
San Lorenzo
Paraguay

Ma. Carolina Zúñiga-López
Universidad de Chile
Santiago
Chile

Ritva Repo-Carrasco-Valencia
Universidad Nacional Agraria la Molina
Lima
Peru

Claudia Monika Haros
Institute of Agrochemistry and Food Technology (IATA-CSIC)
Valencia
Spain

Editorial Office
MDPI AG
Grosspeteranlage 5
4052 Basel, Switzerland

This is a reprint of the Proceedings, published open access by the journal *Biology and Life Sciences Forum* (ISSN 2673-9976), freely accessible at: https://www.mdpi.com/2673-9976/37/1.

For citation purposes, cite each article independently as indicated on the article page online and as indicated below:

Lastname, A.A.; Lastname, B.B. Article Title. *Journal Name* **Year**, *Volume Number*, Page Range.

ISBN 978-3-7258-4175-2 (Hbk)
ISBN 978-3-7258-4176-9 (PDF)
https://doi.org/10.3390/books978-3-7258-4176-9

Contents

About the Editors

Nancy Chasquibol

Nancy Chasquibol, PhD in Chemical Sciences from the Universidad Nacional Mayor de San Marcos, Peru. She is a researcher and Coordinator of the Grupo de Investigación en Alimentos Funcionales of the Instituto de Investigación Científica (IDIC), Responsible of the Laboratorio de Alimentos Funcionales, and the Principal teacher of Industrial Engineering Career at the University of Lima-Peru. Her research projects are focused on the valorization of Peruvian biodiversity, mainly on ancestral crops to promote the development and innovation in functional foods with high protein content, amino acid profiles, high antioxidants and omega-3, techno-functional properties, tasty, low in saturated fat and sugar, safe and sustainable to contribute to the reduction in Noncommunicable diseases (NCDs) and food insecurity (FAO, SDG 2), by applying clean technology in cooperation with national and international organizations. She is a member of the IA ValSe-Food Network, (Ibero-American Valuable Seeds).

Norma Sammán

Norma Sammán is Professor Emeritus of the Faculty of Engineering of the National University of Jujuy. She belongs to CIITED, an Executive Unit with a double dependency between UNJu and CONICET. She is the director of GIDANO (Food Research and Development Group of Northwest Argentina). Dr. Sammán carries out her activities in Food Science and Technology and Nutrition: chemical, nutritional, and functional characterization of Andean crops, product development, and is the author of numerous scientific publications. She was President of LATINFOODS (Latino-American Food Composition Network) for two periods and was in charge of the Direction of Jujuy Research and Transfer Center (CIT Jujuy)-UNJu-CONICET (2013–2018). Is a member of the Regional Academic Committee of the Doctorate in Food Science and Technology, and directed doctoral and postdoctoral scholarships. She belongs to the Ia VALSE and Chia Link networks.

Pedro Maldonado

Pedro Maldonado has a PhD in Biochemistry, Chemistry and Food Technology of Montpellier University, France. He is a Tenured Associate Professor at the Faculty of Chemistry and Agroindustry in Escuela Politécnica Nacional (EPN) of Ecuador and has worked at the Department of Food Science and Biotechnology since 2014. Also, Dr. Maldonado is guest lecturer in other Ecuadorian universities. Currently, he is the general coordinator of the Ecuadorian network of cereals, pseudocereals, tubers, roots, and legumes. His work focused on the valorization of Ecuadorian farinaceous products and by-products, in particular tubers, pseudocereals and pulses with high biological value to develop new functional foods with high nutritional, structural and sensory properties. He has also worked in food safety related to heavy metal analysis of Ecuadorian foods, as well as in the extraction of molecules of interest from non-studied endemical resources from this country.

Laura Mereles

Laura Mereles has a PhD in Food Sciences from the University National of Asunción, Paraguay (UNA). She has developed her professional work in the Department of Food Biochemistry of the Faculty of Chemical Sciences (FCQ-UNA) and other Universities and Foreign Research Centers since 2015. Currently, she is Principal Investigator of the Biodiversity, Food and Health Group (BIOALSA), Department of Food Biochemistry of the Directorate of Research (FCQ-UNA). She is a member of the Chia-Link Network, and the ValSe-Food Network (Valuable Ibero-American Seeds) representing the Paraguay group. Her work as Professor Researcher in Food Sciences is focused on the Food Safety, Public Health and Production of Healthy Foods areas, within the framework of environmental sustainability, through research lines such as the study of food composition with emphasis on seeds, grains, fats and oils, bioprospecting of under-utilized regional food resources with antioxidant potential, and food safety and development of methodologies for the evaluation of toxins in food. e-mail: lauramereles@qui.una.py

Ma. Carolina Zúñiga-López

Ma. Carolina Zúñiga-López has a chemical undergraduate degree specializing in Industrial Chemistry from the Pontificia Universidad Católica de Chile, she later completed a PhD in Analytical Chemistry and two post-doctorates in Chemical Engineering. She is currently an Associate Professor in the Department of Inorganic and Analytical Chemistry of the Faculty of Chemical and Pharmaceutical Sciences of the University of Chile. Her research has focused on analytical chemistry as a tool for determining bioactive compounds extracted from natural products, focusing on evaluating their antioxidant capacity chemical and cellular in vitro and its in vivo effect. She has also worked in Medicinal Chemistry, focused on drug synthesis for Chagas disease (*Trypanosoma cruzi*).

Ritva Repo-Carrasco-Valencia

Ritva Repo-Carrasco-Valencia has a PhD in Food Chemistry from University of Turku, Finland. She has MSc in Food Chemistry and Technology form University of Helsinki, Finland. She is a principal professor at the Food Industry Faculty of the National Agrarian University la Molina (UNALM) in Lima, Peru. She has long time experience in working with Andean native crops, such as quinoa, kañiwa, amaranth, and tarwi (Andean lupin). Her work consists of research on nutritive value and uses of Andean native crops. She is the director of CIINCA, Center of research on Andean native crops of UNALM.

Claudia Monika Haros

Claudia Monika Haros has a PhD in Chemical Sciences from the University of Buenos Aires, Argentina. She has developed her professional work in Universities and National and Foreign Research Centers since 1990. Currently, she is the Head of the Cereals Group of the Institute of Agrochemistry and Food Technology belonging to the Higher Council for Scientific Research of Spain (IATA- CSIC). She is a founding member of the Chia-Link Network, which is included in the ValSe-Food Network (Iberoamerican Valuable Seeds or Valiosas Semillas Iberoamericanas) that she coordinates. Her research on ancestral Latin American crops aims to work in an environment of cooperation with the international scientific community and unite efforts to promote safe, sustainable, tasty, nutritious, and healthy foods with cereals/pseudocereals or their by-products, among other crops such as chia, through collaborations between the sectors involved in research, academic institutions, industry, and society.

Preface

Within this reprint, readers will find a compilation of recent investigations conducted by the ValSe-Food Group on valuable seeds and other crops such as: amaranth (*Amaranthus* spp.); quinoa (*Chenopodium quinoa*); kañiwa (*Chenopodium pallidicaule*); chia (*Salvia hispanica* L.); peanut (*Arachis hypogaea*); Andean maize (*Zea mays* L.); purple corn morado (*Zea mays* L.); tarwi (*Lupinus mutabilis*); broad beans (*Vicia faba* L.); bean (*Phaseolus vulgaris* L.); indian bean (*Cynophalla retusa*); Mistol (*Sarcomphalus mistol*); moringa (*Moringa oleifera*); kurugua (*Sicana odorifera*); wild chili (*Capsicum chacoense*), white carob tree (*Neltuma alba*), Paraguayan coconut (*Acrocomia aculeata*); native potatoes (*Solanum andigenum*); mashua negra (*Tropaeolum tuberosum*). The reprint provides thorough and current insights into multiple aspects of food science, focusing on production, use, structural properties, and chemical makeup. It emphasizes carbohydrates, fibers, bioactive compounds, proteins, and lipids found in seeds and other plant components. The material explores the processes involved, the range of food products and applications, as well as the nutritional and health impacts linked to these crops.

Nancy Chasquibol, Norma Sammán, Pedro Maldonado, Laura Mereles, Ma. Carolina Zúñiga-López, and Ritva Repo-Carrasco-Valencia
Volume Editors

biology and life sciences forum

MDPI

Editorial

Preface of the VI International Congress la ValSe-Food

Nancy Chasquibol [1], Norma Sammán [2], Pedro Maldonado [3], Laura Mereles [4], Ma. Carolina Zúñiga-López [5], Ritva Repo-Carrasco-Valencia [6] and Claudia Mónika Haros [7,*]

1 Ingeniería Industrial, Grupo de Investigación en Alimentos Funcionales, Instituto de Investigación Científica, Universidad de Lima, Av. Javier Prado Este 4600, Fundo Monterrico Chico, Surco, Lima 15023, Peru; nchasquibol@ulima.edu.pe

2 Facultad de Ingeniería, CIITED-Universidad Nacional de Jujuy-CONICET, San Salvador de Jujuy 4600, Argentina; normasamman@gmail.com

3 Department of Food Science and Biotechnology, Escuela Politécnica Nacional, Quito 17-01-2759, Ecuador; pedro.maldonado@epn.edu.ec

4 Bioquímica de Alimentos Department, BIOALSA Group, Facultad de Ciencias Químicas, Universidad Nacional de Asunción (FCQ-UNA), San Lorenzo P.O. Box 1055, Paraguay; laurameréles@qui.una.py

5 Departamento de Química Inorganica y Analitica, Facultad de Ciencias Químicas y Far-Maceuticas, Universidad de Chile, Santiago 8380492, Chile; mczuniga@ciq.uchile.cl

6 Centro de Investigación e Innovación en Granos Andinos (CIINCA), Universidad Nacional Agraria la Molina, Lima 012, Peru; ritva@lamolina.edu.pe

7 Cereal Group, Instituto de Agroquímica y Tecnología de Alimentos, Spanish Council for Scientific Research (IATA-CSIC), 46980 Valencia, Spain

* Correspondence: cmharos@iata.csic.es; Tel.: +34-963900022

1. Introduction

The growing interest in Ibero-American Valuable Seeds has led to intensified research efforts in recent decades. This book encapsulates the proceedings of the VI International Ia ValSe-Food Network, titled "Promoting biodiversity, sustainability, and food security through Ibero-American ancestral crops", which took place at the Universidad de Lima from September 23 to 25 in Lima, Peru.

Ibero-American crops are currently underutilized, with cultivation levels remaining relatively low. Nonetheless, there has been a significant rise in global interest in these crops, resulting in increased production and market prices. These seeds have gained recognition from food scientists and producers for their outstanding nutritional benefits. They are rich in high-quality proteins, fibers, and starches, have distinctive properties, and also have a wealth of micronutrients including vitamins, minerals, and bioactive compounds.

The network's focus on studying and promoting Ibero-American crops aims to support the conservation of genetic diversity. Ibero-American crops, with their unique traits developed through centuries of natural selection, are essential resources for future breeding programs and for creating crops that are resilient to climate change. By preserving and utilizing these crops, we can help maintain biodiversity and safeguard valuable genetic resources against environmental challenges.

The primary goal of the network is to foster a collaborative environment among the global scientific community to advance the development of safe, sustainable, and nutritious food from ancient crops. This involves partnerships among researchers, institutions, industry professionals, and the public. Additionally, the International ValSe-Food Network seeks to promote the sustainable progress of science and technology through the study and application of Latin-American crops.

2. Topics

The conference was structured into different sections to cover various topics related to agriculture, food technology, nutrition, health promotion, and climate change.

- Session I: Agronomy and Crop Diversity in the Face of Climate Change.
- Session II: Food Technology and Innovation. Biodiversity and Trends. Sustainable Management of Food Waste.
- Session III: Nutritional and Health Promotion. Political and Socioeconomic Perspectives.

Each section featured keynote presentations, research paper presentations, panel discussions, interactive poster sessions, and concluded with roundtable discussions to further facilitate knowledge exchange and collaboration among participants. The conference covered a wide range of topics, fostering interdisciplinary dialog and promoting holistic approaches to sustainable agriculture, food technology, nutrition, health promotion, and climate change.

3. Committees

3.1. Scientific Committee

- Nancy Chasquibol

Ingeniería Industrial, Grupo de Investigación en Alimentos Funcionales, Instituto de Investigación Científica, Universidad de Lima, Lima, Peru

- Norma Sammán

Facultad de Ingeniería, CIITED-Universidad Nacional de Jujuy-CONICET, Jujuy, Argentina

- Pedro Maldonado

Department of Food Science and Biotechnology, Escuela Politécnica Nacional, Quito, Ecuador

- Laura Mereles

Bioquímica de Alimentos Department, BIOALSA Group, Facultad de Ciencias Químicas, Universidad Na-cional de Asunción (FCQ-UNA), San Lorenzo, Paraguay

- Ma. Carolina Zúñiga-López

Departamento de Química Inorganica y Analitica, Facultad de Ciencias Químicas y Farmaceuticas, Universidad de Chile, Santiago, Chile

- Ritva Repo-Carrasco Valencia

Centro de Investigación e Innovación en Granos Andinos (CIINCA), Universidad Nacional Agraria la Molina, Lima, Peru

- Claudia Mónika Haros

Coordinator of International ValSe-Food Network
Cereal Group, Instituto de Agroquímica y Tecnología de Alimentos, Spanish Council for Scientific Research (IATA-CSIC), Valencia, Spain

3.2. Organizer Committee-Completer

- Nancy Chasquibol

Ingeniería Industrial, Grupo de Investigación en Alimentos Funcionales, Instituto de Investigación Científica, Universidad de Lima, Lima, Peru

- Claudia Mónika Haros

Coordinator of International ValSe-Food Network
Cereal Group, Instituto de Agroquímica y Tecnología de Alimentos, Spanish Council for Scientific Research (IATA-CSIC), Valencia, Spain

Funding: This congress was funded by Universidad de Lima and ValSe-Food Network.

Conflicts of Interest: The authors declare no conflicts of interest.

biology and life sciences forum

MDPI

Editorial

Statement of Peer Review

Nancy Chasquibol [1] and Claudia M. Haros [2,*]

[1] Ingeniería Industrial, Grupo de Investigación en Alimentos Funcionales, Instituto de Investigación Científica, Universidad de Lima, Av. Javier Prado Este 4600, Fundo Monterrico Chico, Surco, Lima 15023, Peru; nchasquibol@ulima.edu.pe

[2] Cereal Group, Instituto de Agroquímica y Tecnología de Alimentos, Spanish Council for Scientific Research (IATA-CSIC), 46980 Valencia, Spain

* Correspondence: cmharos@iata.csic.es; Tel.: +34-963900022

In submitting conference proceedings to *Biology and Life Sciences Forum*, the volume editors of the proceedings certify to the publisher that all papers published in this volume have been subjected to peer review administered by the volume editors. Reviews were conducted by expert referees to the professional and scientific standards expected of a proceedings journal.

- Type of peer review: single-blind; double-blind; triple-blind; open; other (please describe): open; the editors
- Conference submission management system: the conference was organized through the web: https://civf.ulima.edu.pe/
- Number of submissions sent for review: 24
- Number of submissions accepted: 24
- Acceptance rate (number of submissions accepted/number of submissions received): all accepted
- Average number of reviews per paper: two or three
- Total number of reviewers involved: seven
- Any additional information on the review process: no comments

Reviewers' Criteria: The editors are the reviewers and are provided with guidelines and criteria by the conference organizers. These criteria include assessing the significance of the research, the rigor of the methodology, the clarity of presentation, the originality of the work, and the relevance to the field. Reviewers also comment on the strengths and weaknesses of the manuscript and may suggest improvements or revisions.

Editorial Decision: based on the feedback from peer reviewers, the editors make a decision. The possible decisions include the following:

- Acceptance: the manuscript is accepted for publication without major revisions.
- Minor Revisions: the manuscript needs minor changes or clarifications before acceptance.
- Major Revisions: the manuscript requires substantial revisions and needs to be reevaluated after revisions.
- Rejection: the manuscript does not meet the necessary quality and is not suitable for publication in its current form.

Revisions and Resubmission: If revisions are requested, the authors make the necessary changes to address the editors' comments and concerns. The revised manuscript is then resubmitted to the editors for further evaluation.

Final Decision: The Editorial or handling editor makes a final decision based on the revised manuscript and feedback from the authors. This decision may lead to acceptance, further revisions, or rejection.

Conflicts of Interest: The author declares no conflicts of interest.

biology and life sciences forum

MDPI

Proceeding Paper

Development of Food Hydrogels with Andean Purple Corn (*Zea mays* L.) Extracts and Cushuro (*Nostoc sphaericum*) Polysaccharide: Rheological Characterization †

Cecilia A. Arenas [1,*] and Nancy A. Chasquibol [2]

1 Carrera de Ingeniería Industrial, Universidad de Lima, Av. Javier Prado Este 4600, Surco, Lima 15023, Peru
2 Ingeniería Industrial, Grupo de Investigación en Alimentos Funcionales, Instituto de Investigación Científica, Universidad de Lima, Av. Javier Prado Este 4600, Fundo Monterrico Chico, Surco, Lima 15023, Peru; nchasquibol@ulima.edu.pe
* Correspondence: 20182263@aloe.ulima.edu.pe
† Presented at the VI International Congress la ValSe-Food, Lima, Peru, 23–25 September 2024.

Abstract: Andean purple corn (*Zea mays* L.) is an ancient native Peruvian crop that is currently used in Peruvian gastronomy. Cushuro is a cyanobacteria from the Andean lakes of Peru. They have considerable amounts of bioactive compounds that can improve the physicochemical properties of foods. The objective of this research was to characterize the rheological and functional properties of food hydrogels developed with purple corn extracts, red prickly pear fruit pulp, and cushuro polysaccharide (CP). Acid-soluble polysaccharides obtained from the *Nostoc sphaericum* variety from Ancash, Peru, as well as Peruvian purple corn extracts, were used. Food hydrogels at concentrations ranging from 0.5% to 3.5% (w/v) were elaborated by dispersing the polysaccharides in a 4:1 extract/pulp (v/v) ratio. Likewise, control samples with Tara (*Caesalpinia spinosa*) gum (TG) were made. The effect of hydrocolloid concentration (0.5; 1.5; 2.5; 3.5%) on the rheological properties was evaluated using a unifactorial design. CP and TG hydrogels exhibited a shear-thinning nature, a concentration-dependent yield point (0.02–29.91 Pa; 2.01–508.39 Pa), and high antioxidant activity and phenolic content. Adding CP revealed slow structural regeneration, while TG showed thixotropy and a symmetric hysteresis loop. CP gels showed a fluid-like structure with viscoelastic properties ($G'' > G'$) even in the highest concentration evaluated (3.5%), contrary to TG gel that had a more solid (gel-like) structure ($G' > G''$) at a low concentration (1.5%). These results showed a suitable rheological profile and desirable properties of the food hydrogels development for the functional food industry and processing.

Keywords: bioactive compounds; cushuro; food hydrogels; polysaccharide; purple corn; rheology

1. Introduction

Andean purple corn (*Zea mays* L.) is an ancient native Peruvian crop that is currently used in Peruvian gastronomy. The intense purple color on its pericarp and cob and, consequently, on its extracts is used to prepare drinks like "chicha morada" and desserts like "mazamorra morada". According to Salvador-Reyes et al. [1], this variety of corn is a potential ingredient for the development of new products as it can provide a variety of colors, flavors, and textures, elevated phenolic content, and high antioxidant activity, contributing to a reduction in the use of additives in the food industry while providing health-promoting effects.

Cushuro (*Nostoc sphaericum*) is a bluish-green type of cyanobacteria found in the Andean lakes of Peru. Among its beneficial bioactive compounds, such as protein (28.18 ± 0.33%), iron (4.76 ± 0.08 mg/100 g), and calcium (377.80 ± 1.43 mg/100 g), it contains a significant amount of polysaccharides [2] which have thickening and structural properties as well as potential health benefits (e.g., antitumor, anticoagulant) [3]. Tara (*Caesalpinia spinosa*) gum, also known as Peruvian carob, is a natural hydrocolloid obtained by grinding the endosperm of the seeds of *Caesalpinia spinosa*. It has been approved as a food additive by the Food Chemicals Codex and primarily functions as a thickener and stabilizer [4]. Polysaccharide hydrogels are commonly used in food products due to their ability to modify the rheology of complex food systems, perform a range of technological functions, and create innovative products to answer consumer's needs.

The objective of this investigation was to develop food hydrogels with cushuro (*Nostoc sphaericum*) polysaccharides and Andean purple corn (*Zea mays* L.) extracts to leverage their multiple properties, gain a clearer understanding, expand their potential applications, and promote the use of ancestral crops and new hydrocolloid sources in the functional food industry and food processing. Therefore, this work evaluates and summarizes the dynamic viscoelastic properties, flow behavior, thixotropy, hysteresis, phenolic content, and antioxidant activity, among other aspects of the rheological and functional profile.

2. Materials and Methods

2.1. Materials and Sample Preparation

The cushuro was obtained from the Cotaparaco district, province of Recuay, Ancash region, Peru. It was dried at 60 °C for 18 h and then pulverized. Cushuro polysaccharide (CP) was extracted following the methodology described by Chasquibol et al. [3] at the Functional Food Laboratory of the University of Lima, Peru. Tara gum (TG) was provided by Molinos Asociados SAC (Lima, Peru). Andean purple corn and red prickly pear fruit were purchased from a local market in Lima, Peru. Purple corn extracts were prepared, taking into consideration the outlines recently reported by Jing et al. [5]. The crushed and milled purple corn cob and whole kernels were added to deionized water (10 mL/g; mL solvent/g sample), treated at 75 °C for 1 h, and then strained to obtain the extract. Red prickly pear fruit was washed, peeled, blended, and strained to obtain the seedless pulp.

Table 1 shows eight formulations developed with Andean purple corn (*Zea mays* L.) extracts, red prickly pear fruit pulp, and two different hydrocolloids: cushuro (*Nostoc sphaericum)* polysaccharide and Tara (*Caesalpinia spinosa*) gum. Food hydrogels at concentrations ranging from 0.5% to 3.5% (w/v) were elaborated by dispersing the hydrocolloids in Andean purple corn extracts with continuous stirring for 1 h at room temperature (approx. 25 °C). The solutions were then heated in a water bath at 80 °C with continuous stirring until complete dissolution and homogenization. After cooling the solutions to 50 °C, prickly pear fruit pulp was added. The hydrogels were strained to remove lumps and stored at 4 °C in 50 mL tubes for further analysis. Purple corn extracts and red prickly pear fruit pulp were combined in a 4:1 (extract/pulp v/v) ratio.

Table 1. Formulations of food hydrogels with Andean purple corn (*Zea mays* L.) extracts, cushuro (*Nostoc sphaericum*) polysaccharide, and Tara (*Caesalpinia spinosa*) gum.

Ingredient	CP0.5	CP1.5	CP2.5	CP3.5	TG0.5	TG1.5	TG2.5	TG3.5
Cushuro polysaccharide (% w/v)	0.5	1.5	2.5	3.5	–	–	–	–
Tara gum (% w/v)	–	–	–	–	0.5	1.5	2.5	3.5
Purple corn extract (% v/v)	80	80	80	80	80	80	80	80
Red prickly pear fruit pulp (% v/v)	20	20	20	20	20	20	20	20

CP: cushuro polysaccharide; TG: Tara gum.

2.2. Rheological Measurements of Food Hydrogels

2.2.1. Steady Shear Properties: Flow Behavior, Thixotropy, and Hysteresis Tests

Rheological measurements were performed 24 h after sample preparation using an MCR92 (Anton Paar GmbH, Graz, Austria) rheometer equipped with a parallel plate geometry (50 mm diameter) and a 1 mm measuring gap. The RheoCompass software (version 1.13) integrated into the rheometer was used for data collection, basic analysis, and visualization of the results. For flow behavior tests, the samples were continuously sheared at rates ranging from 0.1 to 100 s^{-1}. The shear rate-dependent Viscosity Ratio (VR) was determined by calculating the ratio between the viscosity of a sample at a high (ηB) and at a low (ηA) rotational speed (Equation (1)):

$$VR = \eta B / \eta A \quad (1)$$

If the value of VR = 1, the sample shows Newtonian flow behavior. If VR < 1, the sample shows speed-dependent shear-thinning flow behavior, and if VR > 1, the sample shows speed-dependent shear-thickening flow behavior [6].

A 3-interval thixotropy test (3ITT) was conducted to simulate the conditions of application processes. The shear rate was maintained at 0.25 s^{-1} for 90 s (1st interval), at 1000 s^{-1} for 60 s (2nd interval), and then gradually decreased from 1000 to 0.25 s^{-1} over 100 s (3rd interval).

Hydrogels were also subjected to a hysteresis loop test to evaluate their response to increase–decline cycles and assess the reversibility of viscosity. The shear rate increased from 5 to 131 s^{-1} over 90 s and gradually declined back to 5 s^{-1} after being held at 131 s^{-1} for 3 min. All steady-state tests were performed at 25 °C.

2.2.2. Dynamic Viscoelastic Properties: Amplitude and Frequency Sweeps

An oscillatory amplitude sweep test was conducted at an angular frequency of ω = 10 rad/s (1.6 Hz) with strain ranging from 1% to 100% to establish the linear viscoelastic (LVE) region. Subsequently, frequency sweeps from 0.1 to 100 Hz (0.63 to 628.3 rad/s) were performed using an appropriate strain within the LVE region to measure the storage modulus (G'), loss modulus (G''), and complex viscosity (η^*). All dynamic tests were conducted at 25 °C.

2.3. Physicochemical Characterization

2.3.1. Determination of Antioxidant Activity via Radical 2,2-Diphenyl-1-picrylhydrazyl (DPPH)

The antioxidant activity of samples was determined by the DPPH method, with some modifications, at 517 nm with a spectrophotometer (UV-1280 UV-VIS Spectrophotometer Shimadzu, Kyoto, Japan). The results were expressed as mg of Trolox equivalent (TE)/g sample [3].

2.3.2. Total Phenolic Content (TPC)

The total phenolic content (TPC) of the samples was determined using the Folin–Ciocalteu method at 760 nm with a spectrophotometer (UV-1280 UV-VIS Spectrophotometer Shimadzu, Kyoto, Japan). The results were expressed as mg of gallic acid equivalent (GAE)/g sample [3].

2.4. Statistical Analysis

Physicochemical analyses were performed in triplicate, and results were expressed as mean values ± standard deviation. A one-way analysis of variance (ANOVA) was conducted to analyze data at a 95% significance level using Minitab 19.0 Software (Minitab Inc., State College, Palo Alto, CA, USA).

3. Results and Discussion

3.1. Rheological Evaluation

3.1.1. Flow and Viscosity Behavior

Table 2 and Figure 1 present the steady shear flow behavior of CP and TG food hydrogels. For each polysaccharide, all tested solutions (0.5%, 1.5%, 2.5%, 3.5%) exhibited a shear-thinning behavior with a Viscosity Ratio (VR) consistently less than 1 across shear rates ranging from 0 to 100 s^{-1}. All solutions presented a concentration dependence of viscosity, which increased with higher hydrocolloid concentrations.

Table 2. Steady shear properties (Yield point [Pa], Viscosity [mPa·s] at 7.5 s^{-1} and 100 s^{-1} shear rates and Viscosity Ratio [dimensionless]) of food hydrogels with Andean purple corn (*Zea mays* L.) extracts, cushuro (*Nostoc sphaericum*) polysaccharide, and Tara (*Caesalpinia spinosa*) gum.

Parameter	CP0.5	CP1.5	CP2.5	CP3.5	TG0.5	TG1.5	TG2.5	TG3.5
Yield point [Pa] [a]	0.026	0.336	6.322	29.914	2.007	91.979	268	508.39
Viscosity [mPa·s] (shear rate 7.5 s^{-1})	7.52	75.07	883.81	7060.4	303.33	8885.9	41,017	77,550
Viscosity [mPa·s] (shear rate 100 s^{-1})	6.76	47.93	289.49	1071.9	108.27	1233.7	3880.5	4663.6
Viscosity Ratio [b]	0.90	0.64	0.33	0.15	0.36	0.14	0.09	0.06

CP: cushuro polysaccharide; TG: Tara gum. [a] According to the Bingham mathematical regression model. [b] Obtained by dividing viscosity at a shear rate of 100 s^{-1} by viscosity at a shear rate of 7.5 s^{-1}.

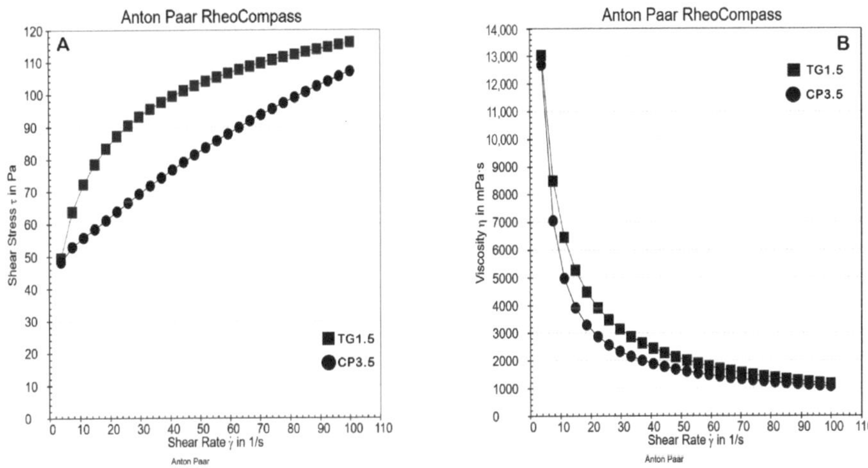

Figure 1. Flow (**A**) and viscosity (**B**) curves of CP3.5 and TG1.5 hydrogels. CP: cushuro polysaccharide; TG: Tara gum.

3.1.2. Yield Point Determination

Typical analysis procedures based on the Bingham mathematical regression model were used to determine the yield point, which is the minimum force required to initiate flow in a sample. The results indicated that CP hydrogels have a relatively low stress threshold before they start to deform (0.026–29.91 Pa), whereas TG hydrogels exhibit higher mechanical strength (2.007–508.39 Pa) even at low concentrations. In application environments (e.g., transport, storage, mixing), CP gels, with their lower yield points, may be more suitable where softer and more flexible materials are needed due to the nature of their internal structural forces. In contrast, TG hydrogels, with their higher yield points, are more appropriate for scenarios requiring greater stress resistance or robust mechanical properties.

3.1.3. Thixotropy Testing and Hysteresis Loop Test

Figure 2 shows the thixotropic properties and hysteresis curves of food hydrogels. In food testing, evaluating structural regeneration is crucial for determining whether the product meets quality standards after processing (e.g., spreading, pouring, or squeezing out of a container). After being subjected to a shear rate of 1000 s^{-1} for 1 min, TG hydrogels showed rapid recovery and structural regeneration. These results are consistent with those reported by Wu et al. [3]. As for TG1.5 gel, the sample achieved complete regeneration of viscosity (100%)—compared to the structural strength at the end of the first interval—90 s after the high shear load interval ended. In contrast, CP3.5 gel remained thinner than its initial state and exhibited considerably slower structural regeneration (32.7%), with regeneration not yet complete even 90 s after the end of the high shear load interval.

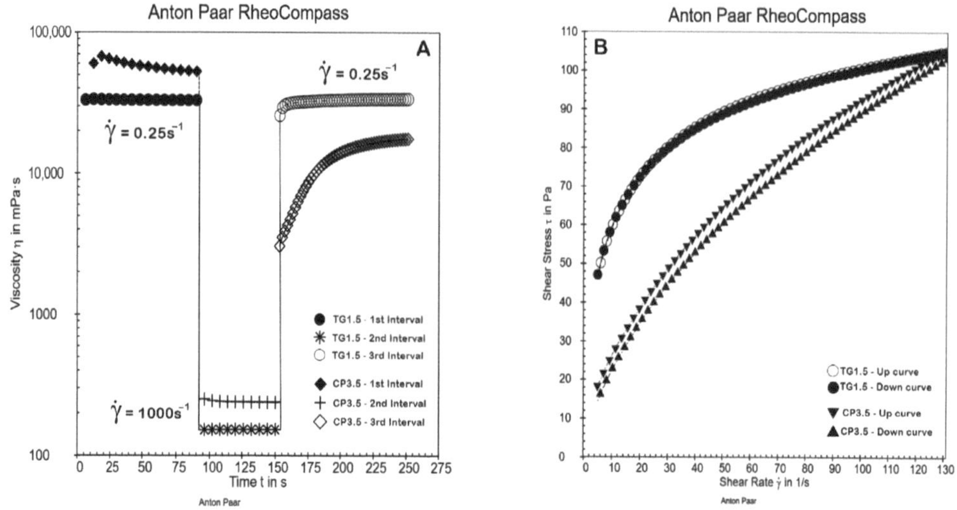

Figure 2. Thixotropy curves (**A**) and hysteresis loop (**B**) of CP3.5 and TG1.5 hydrogels. CP: cushuro polysaccharide; TG: Tara gum.

The results of the hysteresis test illustrate the differences in viscosity between the two types of hydrogels and show how the shear stress varies with increasing and, subsequently, decreasing shear rates. TG1.5 gel sample exhibited a nearly symmetrical hysteresis loop, where the upper curve closely overlapped with the downward curve, displaying an area of 39.86 Pa/s between them. In contrast, CP3.5 hydrogel demonstrated significant structural breakdown, with a hysteresis area of 498.93 Pa/s. For CP hydrogels, their structure does not fully recover and remains thinner than initially, while TG hydrogels achieve complete recovery after the defined rest period.

3.1.4. Dynamic Viscoelastic Properties

Frequency sweeps characterize a sample's time-dependent behavior within the Linear Viscoelastic (LVE) region, where the sample's structure is not destroyed. By evaluating the curve of the storage modulus (G′), the limit of the LVE region for TG1.5 hydrogel was found to be 8.46% shear strain, beyond which G′ function decreases continuously, indicating a gradual breakdown of the superstructure. Under the applied measurement conditions, CP3.5 hydrogel (with G″ > G′) exhibits a fluid-like structure and can be classified as a viscoelastic liquid, with no distinct transition to a region where the viscoelastic properties change significantly (Figure 3).

Figure 4 shows the mechanical spectra of Tara gum solutions at 25 °C. The results of frequency sweeps for TG and CP food hydrogels indicate a decrease in complex viscosity with increasing frequency (shear-thinning behavior), consistent with the steady shear results. For the TG1.5 gel, the loss modulus (G″) was higher than the storage modulus (G′) at low frequencies, indicating that the viscous component of the viscoelastic behavior dominates and describes the liquid state of the sample. This trend continues until a crossover point (sol/gel transition) at 3.53 rad/s, after which G′ becomes predominant at higher frequencies, demonstrating elastic behavior and strong interaction forces. In contrast, CP3.5 hydrogel exhibited a higher loss modulus than storage modulus (G″ > G′) throughout the frequency range, indicating a viscoelastic behavior characterized by mostly unlinked individual molecules with some degree of entanglement [4].

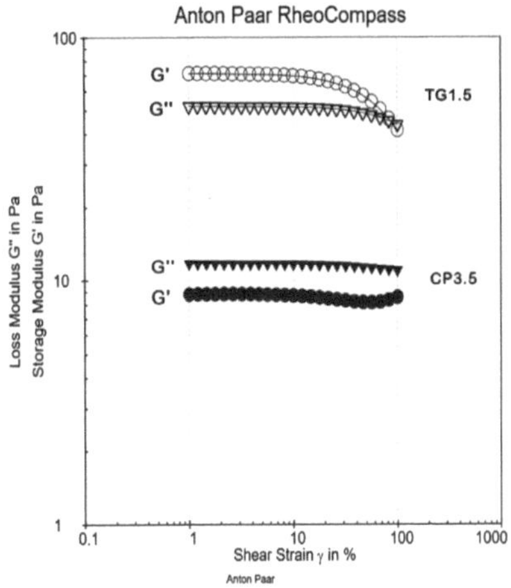

Figure 3. Amplitude sweeps of CP3.5 and TG1.5 hydrogels.

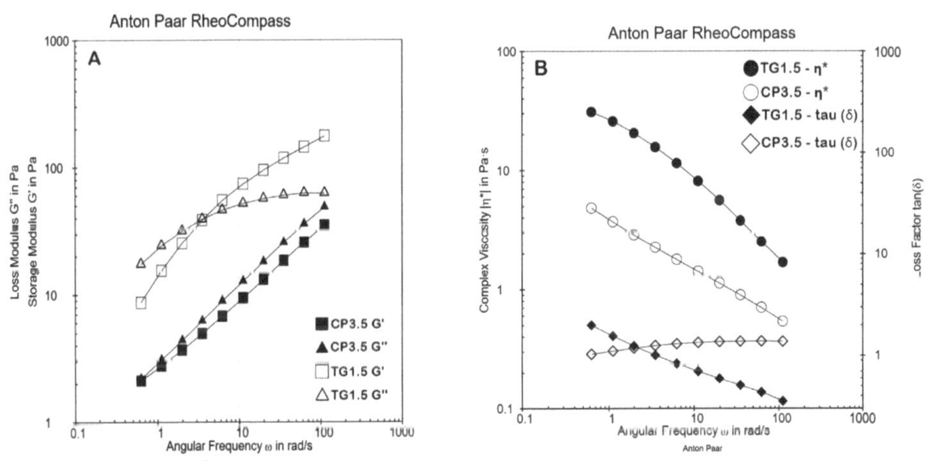

Figure 4. Frequency sweeps (**A**) and complex viscosity (**B**) curves of CP3.5 and TG1.5 hydrogels. CP: cushuro polysaccharide; TG: Tara gum.

3.2. Physicochemical Evaluation

The addition of purple corn extracts, prickly pear fruit pulp, and polysaccharides resulted in significant levels of phenolic compounds and substantial antioxidant capacity. CP3.5 food hydrogel exhibited slightly higher total phenolic content (2.61 ± 0.17 mg GAE/g sample) and antioxidant activity (4.28 ± 0.29 mg TE/g sample) compared to TG3.5 food hydrogel, which had 2.43 ± 0.39 mg GAE/g sample and 4.07 ± 0.22 mg TE/g sample. These findings are consistent with reports by Madhujith et al. [7] and Salvador-Reyes et al. [1], suggesting that these ingredients possess excellent antioxidant properties and may offer health-promoting effects. For the specific analysis of this study, the results of the evaluated formulations were obtained through the experimental pairs design. From the rheological point of view, the results obtained for the CP3.5 and TG1.5 samples are very similar and, consequently, present a better dynamic interpretation of the effects of the studied variable. From the physicochemical point of view, analogous samples with the same concentration of CP and TG presented similar results between both hydrocolloids, so it is concluded that the antioxidant activity and phenolic compounds content are directly related to the amount of raw materials used, as seen for CP3.5 and TG3.5 food hydrogels' evaluation.

4. Conclusions

The hydrocolloid concentrations significantly influenced the rheological properties of the formulated food hydrogels. Specifically, this variable led to increased values of viscosity, pseudoplasticity, and yield point in all food hydrogels. At the highest concentration evaluated, cushuro polysaccharide gels exhibited a dominant viscous component in the viscoelastic behavior, while Tara gum gels displayed elastic behavior and strong interaction forces even at low concentrations. Furthermore, the integration of ancestral crop extracts with novel polysaccharide sources enhanced the total phenolic content and antioxidant activity of the hydrogels. These results highlight the potential for these hydrogels to be effectively used in various applications or to innovate food products leveraging traditional ingredients and bioactive compounds.

Author Contributions: Conceptualization, C.A.A. and N.A.C.; Methodology, C.A.A. and N.A.C.; Investigation and Data analysis, C.A.A.; Writing—original draft preparation, C.A.A.; writing—review and editing, C.A.A. and N.A.C. All authors have read and agreed to the published version of the manuscript.

Funding: This research received no external funding.

Institutional Review Board Statement: Not applicable.

Informed Consent Statement: Not applicable.

Data Availability Statement: Data is contained within the manuscript.

Acknowledgments: This work was supported and developed at Laboratorio de Alimentos Funcionales de la Carrera de Ingeniería Industrial, Universidad de Lima, Peru.

Conflicts of Interest: The authors declare no conflicts of interest.

References

1. Salvador-Reyes, R.; Pedrosa Silva Clerici, M.T. Peruvian Andean maize: General characteristics, nutritional properties, bioactive compounds, and culinary uses. *Food Res. Int.* **2020**, *130*, 108934. [CrossRef] [PubMed]
2. Méndez, A.S.; Victoriano, P.; Soto, G.H.; Zambrano, C.W.; Marín, M.O.; Zapata, R.J.; Maquera, M.M.; Fernandez, H.R.; González, A.J.; Zuffo, A.; et al. Physicochemical Evaluation of Cushuro (*Nostoc sphaericum Vaucher ex Bornet & Flahault*) in the Region of Moquegua for Food Purposes. *Foods* **2023**, *12*, 1939. [CrossRef] [PubMed]
3. Chasquibol, N.; Sotelo, A.; Tapia, M.; Alarcón, R.; Goycoolea, F.; Perez-Camino, M.d.C. Co-Microencapsulation of Cushuro (*Nostoc sphaericum*) Polysaccharide with Sacha Inchi Oil (*Plukenetia huayllabambana*) and Natural Antioxidant Extracts. *Antioxidants* **2024**, *13*, 680. [CrossRef] [PubMed]
4. Wu, Y.; Ding, W.; Jia, L.; He, Q. The rheological properties of tara gum (*Caesalpinia spinosa*). *Food Chem.* **2015**, *168*, 366–371. [CrossRef] [PubMed]
5. Jing, P.; Giusti, M.M. Effects of Extraction Conditions on Improving the Yield and Quality of an Anthocyanin-Rich Purple Corn (*Zea mays* L.) Color Extract. *J. Food Sci.* **2007**, *72*, C363–C368. [CrossRef] [PubMed]
6. Mezger, T. Basic Principles, in II (Ed.). In *The Rheology Handbook*; Vincentz Network: Hannover, Germany, 2014; pp. 33–35.
7. Madhujith, T.; Wedamulla, N.E.; Gamage, D.A.S. Biological macromolecules as antioxidants. In *Biological Macromolecules*; Academic Press: Cambridge, MA, USA, 2022; pp. 139–164. [CrossRef]

biology and life sciences forum

MDPI

Proceeding Paper

Nutritional Interest of *Geoffroea decorticans* "Chañar": A Native Species from the Province of Mendoza, Argentina †

Pablo Matías Molina, Ana Paz Vignoni, Analía Valdes and Emilia Elisabeth Raimondo *

Facultad de Ciencias Agrarias, Universidad Nacional de Cuyo, Mendoza PC 5505, Argentina;
pmolina@fca.uncu.edu.ar (P.M.M.); avignoni@fca.uncu.edu.ar (A.P.V.); avaldes@fca.uncu.edu.ar (A.V.)
* Correspondence: emilia.raimondo@gmail.com

† Presented at the VI International Congress la ValSe-Food, Lima, Peru, 23–25 September 2024.

Abstract: The "chañar" (*Geoffroea decorticans*) is a tree native to the drylands of South America, historically valued for its nutritional, medicinal, and energy-providing properties. The significance of this species lies in its adaptation to conditions of water and salt stress, as well as its tolerance to wide thermal fluctuations, making it a candidate for utilization in climate change adaptation strategies. This study aimed to quantify the nutritional and mineral contributions of *G. decorticans* fruits from the province of Mendoza, Argentina. Representative specimens were selected, and the fruits were manually harvested for an analysis. The moisture and energy contents were determined using official analytical techniques. The evaluation of the nutritional components was conducted on a dry weight basis, including the total mineral, protein, fat, and total carbohydrate contents. In the mineral fraction test, nitrogen, phosphorus, potassium, sodium, calcium, and magnesium were quantified. The results revealed an adequate protein content (5.27 ± 0.06%) and elevated levels of crude fiber (19.27 ± 0.46%) and total carbohydrates (85.53 ± 0.98%). A high fiber content contributes to satiety, and its consumption could significantly enhance the population's dietary reference intake. Although the mineral profile appears satisfactory, further investigation is required to clarify the factors affecting the bioavailability of each element. Even though there are existing studies on the variation of nutritional properties across different geographic regions, no local studies were identified. This research provides valuable data for the revaluation of ancestral species with nutritional significance, especially considering the growing trend towards the use of native plants in gastronomy.

Keywords: ancestral fruit; *Geoffroea decorticans*; mineral composition; native plants in gastronomy; nutritional profile

1. Introduction

Geoffroea decorticans (Gillies ex Hook. and Arn.) Burkart is a species native to South America, belonging to the botanical family *Fabaceae*. It is commonly found in the drylands of Argentina, northern Chile, Bolivia, southern Peru, the Paraguayan Chaco, and western Uruguay. In Argentina, it has a broad distribution, ranging from Jujuy to the northern Patagonia [1].

It inhabits dry environments with low precipitation, at altitudes ranging from 100 to 2600 m a.s.l., across different edaphic conditions but generally in fine-textured soils, which are often prone to flooding or salinity. Within its natural distribution range, isolated specimens can be observed, or they may form closed groves, supported by its strong gemmiferous roots, which also enable it to colonize new areas. This behavior characterizes it as a pioneering species, crucial for biodiversity conservation [2]. Its adaptation to water and salt stress, as well as large thermal fluctuations, confers productive potential under climate change scenarios. The availability of this species in nurseries specialized in native flora demonstrates its capacity for domestication and cultivation, promoting its propagation while preventing the degradation of natural habitats.

Commonly known as "chañar", this tree has been traditionally and extensively utilized by indigenous peoples and rural inhabitants. Locally, its wood is used in the manufacture of furniture, household items, tool handles, and as fuel due to its energetic properties. The leaves and bark (which has dyeing properties) are employed in traditional medicine. Its fruit contains a sweet and pasty pulp, rich in carbohydrates and fibers, which provides significant nutritional potential and has been used ancestrally both for food and as fodder [3]. It can be consumed fresh or used as an ingredient in various recipes, including sweet foods ("arrope") and alcoholic ferments ("aloja").

The use of pre-Hispanic plants in culinary preparations has recently gained international prominence, highlighting the need for rigorous botanical and nutritional characterization. In this context, the aim of this work was to determine the nutritional and mineral composition of *G. decorticans* fruits from the province of Mendoza, Argentina, as a contribution to the development of local gastronomy.

2. Materials and Methods

2.1. Study Area

Fruit collection was conducted in Montecaseros District, General San Martín Department, Mendoza, Argentina, in an area on the boundary between grapevine crops and natural vegetation [4]. The climate is desertic, with an average annual temperature of 15 °C, average annual precipitation reaching 200 mm during the spring-summer period, and an average altitude of 620 m a.s.l. The vegetation in the natural environment corresponds to the classic Monte Phytogeographic Province, recently classified within the Septentrional District [5], where evergreen shrub steppes and shrublands predominate, along with small relics of forests and grasses in open areas with low tree and shrub cover. The soils are poorly developed Entisols, characterized by a strong moisture deficit for most of the year and predominance of sandy textures, with frequent fixed and semi-fixed dunes and availability of groundwater.

2.2. Botanical Description of G. decorticans

G. decorticans is a tree that can reach up to 10 m in height (Figure 1). Its most distinctive feature is its bark; grayish, it sheds in discontinuous, longitudinal, and irregular plates and strips, revealing the new bright green tissue underneath, which provides photosynthetic activity and a unique appearance (Figure 2). The leaves are pinnate, deciduous, and gray-green, ranging from 1 to 7 cm in length [1]. The flowers are hermaphroditic, measuring 0.8 to 1 cm in length, fragrant, with a striking papilionaceous corolla that is yellow-orange with red stripes. The fruit is a drupaceous legume, measuring 1.5 to 3.5 × 1.5 to 2.4 cm, being ovoid to slightly compressed, glabrous, smooth, and reddish. It generally contains a single seed (exceptionally two), measuring 1.2 to 1.8 cm in length, with a thin reddish testa [2]. In Argentina, flowering occurs in spring, between September and October, depending on the latitude. Fruiting is typically complete during January and February.

Figure 1. General appearance of a chañar tree.

Figure 2. Details of the bark of a chañar tree.

2.3. Sampling

Fruit collection was carried out in the study area at the end of summer 2024. Eighteen representative trees were selected randomly, taking into account their satisfactory phytosanitary condition. The fruits were manually harvested and stored in paper bags until the sample preparation stage, which occurred 24 h later.

2.4. Sample Preparation and Chemical Analysis

The sample preparation began with the separation of the whole fruit (Figure 3) into two fractions, the endocarp and seed (S) and mesocarp and exocarp, hereafter referred to as "pulp" (P) (Figure 4). Both fractions were weighed fresh. The mesocarp and exocarp were dehydrated in a forced-air oven until reaching a constant weight. To carry out the analysis, the dehydrated mesocarp and exocarp were ground using a knife grinder to obtain a homogeneous flour (Figure 5). All determinations were performed in triplicate, following official analytical techniques [6,7].

Figure 3. Whole chañar fruits (AD) from sample A.

Figure 4. Chañar fruits separated into the endocarp and seed (AS) and mesocarp and exocarp (AP) from sample A.

Moisture: Indirect method involving drying in an oven at 70 °C, until a constant weight is achieved (Method of AOAC 167.03 15th Ed. 1990).

Dry material: Method of AOAC 167.03 15th Ed. 1990.

Total fat: Direct method involving extraction with ethyl ether (crude fat) (AOAC 920.39 C 15th Ed. 1990). A Soxhlet gravimetric method was used.

Crude protein: Kjeldahl method (AOAC 984.13 16 Ed. 5th revision 1999), determining nitrogen, using 6.25 as a protein conversion factor.

Ashes: Direct method (AOAC 942.05 15th Ed. 1990) involving incineration in a muffle furnace (at 500 ± 10 °C) until the absence of black spots in the ashes.

Total carbohydrates (including dietary fiber): Determined by difference, using the following formula (Chapter V Argentine Food Code):

$$100 - (\text{weight in grams [protein + fat + water + ash]}), \text{ in 100 g of food}$$

Mineral profile: Determined based on the ashes, involving a chemical analysis. Sodium (AOAC 974.01 18th Ed. 1997), potassium (AOAC 974.01 18th Ed. 1997), phosphorus (AOAC 986.24 18th Ed. 1997), calcium, and magnesium (AOAC 968.31 15th Ed. 1990).

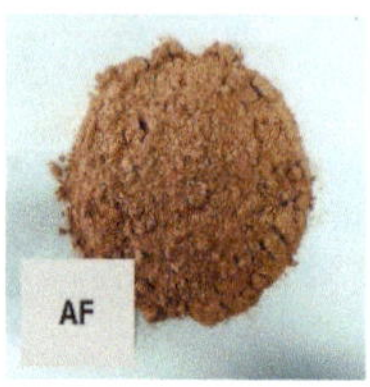

Figure 5. Chañar fruit flour (AF) from sample A.

2.5. Statistical Analysis

All determinations were performed in triplicate (A, B, and C). The results were calculated on both dry and wet bases and are expressed as average values with standard deviations, using the SPSS® statistical software v29.

3. Results and Discussion

3.1. Centesimal Composition of the "Chañar" Fruit

The results obtained are presented in Table 1.

Table 1. Average centesimal composition of chañar fruits.

Nutritional Composition (100 g)	Values Expressed on a Wet Basis	Values Expressed on a Dry Basis
Humidity (g)	15.03 ± 1.72	
Dry matter (g)		84.97 ± 1.72
Ash (g)	6.90 ± 0.21	8.12 ± 0.21
Total protein (g)	4.48 ± 0.10	5.27 ± 0.06
Total fat (g)	0.92 ± 0.05	1.09 ± 0.04
Total carbohydrates (g)	72.67 ± 1.64	85.53 ± 0.98

The moisture contents of the fresh fruit ranged from 13.67% to 16.97%. This characteristic is advantageous in areas without access to electric preservation methods. The values were higher than those reported by Maschio et al. [8] in the Argentine provinces of Catamarca and Río Negro, which may be attributed to climatic or phenological conditions at the time of collection. The protein content was 4.48%, similar to that of the other two regions, while the caloric value was notably lower. Regarding lipids, the values obtained in Mendoza represented 26% and 15% of those found in Catamarca and Río Negro, respectively. The carbohydrate fraction was the most significant in Mendoza samples (85.53 ± 0.98%), as well as in Catamarca (84.90 ± 7.67%) and Río Negro (81.60 ± 5.23%), with very similar absolute and relative values. To make these values comparable, the sum of the carbohydrates and fibers was considered. The results of the present work are in line with the work of Maschio et al. [8]. These results are also consistent with those presented by Orrabalis [9], who notes that "chañar" is an energy-rich fruit.

3.2. Mineral Composition of the "Chañar" Fruit

The mineral profile is presented in Table 2.

Table 2. Average mineral composition of chañar fruits.

Mineral Composition (100 g)	Values Expressed on a Wet Basis	Values Expressed on a Dry Basis
Sodium (mg)	31.1 ± 4.5	36.6 ± 5.8
Potassium (mg)	1103.6 ± 22.9	1298.9 ± 0.8
Calcium (mg)	339.6 ± 7.0	399.7 ± 0.3
Magnesium (mg)	247.8 ± 5.1	291.6 ± 0.2
Phosphorus (mg)	100.6 ± 2.0	118.4 ± 3.0

Costamagna et al. [10] determined a sodium content of 23 mg Na/100 g, which was slightly lower than the value found for Mendoza. However, in both cases, the sodium content is considered low. The mineral contents obtained from chañar trees in northwestern Argentina contrast with those reported in our study, representing 239% of the potassium, 33% of the calcium, 38% of the magnesium, and 2% of the phosphorus found in Mendoza.

4. Conclusions

This study quantitatively determined the nutritional and mineral compositions of *G. decorticans* fruits in Mendoza, Argentina. This provides local information on an innovative aspect such as the nutritional contribution of a native forest species. Laboratory analyses revealed characteristics of the chañar that highlight its significance for the food industry, such as its low moisture content (which supports preservation at room temperature without refrigeration until consumption) and adequate levels of proteins, calcium, and magnesium, given their nutritional importance.

The results indicate the need to further investigate the bioavailability of each chemical element and the variability in the synthesis of nutritional compounds. This includes linking these factors with environmental and climatic aspects and promoting this species to diversify and add value to the local gastronomy industry.

Author Contributions: Conceptualization, P.M.M. and E.E.R.; methodology, A.P.V. and A.V.; software, P.M.M. and A.P.V.; validation, P.M.M., A.P.V. and A.V.; formal analysis, A.V.; investigation, P.M.M. and E.E.R., resources, E.E.R., data curation, P.M.M.; writing—original draft preparation, E.E.R.; writing—review and editing, P.M.M. and A.P.V.; visualization, P.M.M. and A.P.V.; supervision, E.E.R.; project administration, E.E.R.; funding acquisition, E.E.R. All authors have read and agreed to the published version of the manuscript.

Funding: This research was funded by Faculty of Agrarian Sciences National University of Cuyo, Argentina, University of Lima and La ValSe-Food-CYTED (119RT0567).

Institutional Review Board Statement: Not applicable.

Informed Consent Statement: Not applicable.

Data Availability Statement: The article contains all the trial data.

Acknowledgments: This work was supported by grant Ia ValSe Food-CYTED (ref. 119RT0567), Universidad de Lima, Perú; and Facultad de Ciencias Agrarias, Universidad Nacional de Cuyo, Mendoza, Argentina.

Conflicts of Interest: The authors declare no conflict of interest

References

1. Anton, A.M.; Zuloaga, F.O. (Eds.) Flora Argentina. Available online: http://conosur.floraargentina.edu.ar/species/details/9564 (accessed on 5 August 2024).
2. Burkart, A. La posición sistemática del "chañar" y las especies del género *Geoffroea* (*Leguminosae-Dalbergieae*). *Darwiniana* **1949**, 9, 9–23. Available online: https://flora-arborea-chile.linnaeus.naturalis.nl/linnaeus_ng/app/views/literature2/reference.php?id=94319 (accessed on 5 August 2024).

3. Costamagna, M.S.; Zampini, I.C.; Alberto, M.R.; Cuello, S.; Torres, S.; Pérez, J.; Quispe, C.; Schmeda-Hirschmann, G.; Isla, M.I. Polyphenols rich fraction from *Geoffroea decorticans* fruits flour affects key enzymes involved in metabolic syndrome, oxidative stress and inflammatory process. *Food Chem.* **2016**, *190*, 392–402. [CrossRef] [PubMed]
4. Vignoni, A.P.; Peralta, I.E.; Abraham, E.M. Fragmented areas due to agricultural activity: native vegetation dynamics at crop interface (Montecaseros, Mendoza, Argentina). Revista de la Facultad de Ciencias Agrarias. Universidad Nacional de Cuyo. Mendoza. *Argentina* **2023**, *55*, 46–60. [CrossRef]
5. Arana, M.D.; Natale, E.S.; Ferretti, N.E.; Romano, G.M.; Oggero, A.J.; Martínez, G.; Posadas, P.E.; Morrone, J.J. *Esquema Biogeográfico de la República Argentina*; Opera Lilloana: San Miguel de Tucumán, Argentina, 2021; pp. 56–238. Available online: https://www.lillo.org.ar/editorial/index.php/publicaciones/catalog/book/253 (accessed on 7 May 2024).
6. ANMAT. Código Alimentario Argentino. 2024. Available online: https://www.argentina.gob.ar/anmat/codigoalimentario (accessed on 7 May 2024).
7. AOAC. Official Methods of Analysis. 1990. Available online: https://archive.org/stream/gov.law.aoac.methods.1.1990/aoac.methods.1.1990_djvu.txt (accessed on 13 May 2024).
8. Maschio, J.I.; Boeri, P.; Dalzotto, D.; Piñuel, L.; Sharry, S. Caracterización química y nutricional de harina de chañar de diferente distribución geográfica. *Rev. Bionatura* **2021**, *6*, 1563–1568. [CrossRef]
9. Orrabalis, C.J. Aprovechamiento Integral de los Frutos de *Geoffroea decorticans* (Chañar) de la Región Fitogeográfica de la Provincia de Formosa. Ph.D. Thesis, Facultad de Ciencias Exactas y Naturales, Universidad Nacional de Córdoba, Córdoba, Argentina, 2014. Available online: https://rdu.unc.edu.ar/handle/11086/1590 (accessed on 13 June 2024).
10. Costamagna, M.S.; Ordoñez, R.M.; Zampini, I.C.; Sayago, J.E.; Isla, M.I. Caracterización Química de la Harina de *Geoffroea decorticans* (chañar). XI Jornadas de Comunicaciones. Facultad de Ciencias Naturales E I.M.L. 2013. Available online: https://www.conicet.gov.ar/new_scp/detalle.php?keywords=glifosato&id=21963&congresos=yes&detalles=yes&congr_id=2099180 (accessed on 13 June 2024).

biology and life sciences forum

MDPI

Proceeding Paper

In Vitro Digestion of Chia Seed Oil Nanoemulsions †

Luciana Julio [1,*], Greilis Quintero-Gamero [2], Estefanía Guiotto [2] and Vanesa Ixtaina [2,3]

1 Instituto de Investigaciones para la Industria Química (INIQUI) (CONICET Salta-Jujuy, UNSa), Av. Bolivia 5150, Salta 4400, Salta Province, Argentina
2 Centro de Investigación y Desarrollo en Criotecnología de Alimentos (CIDCA) (CONICET La Plata, FCE-UNLP, CICBA), Calle 47 y 116, La Plata 1900, Buenos Aires, Argentina; greilisquintero@quimica.unlp.edu.ar (G.Q.-G.); esnagui@hotmail.com (E.G.); vanesaix@hotmail.com (V.I.)
3 Facultad de Ciencias Agrarias y Forestales, Universidad Nacional de La Plata (FCAyF-UNLP), Calle 60 y 119, La Plata 1900, Buenos Aires, Argentina
* Correspondence: luci_julio86@hotmail.com
† Presented at the VI International Congress la ValSe-Food, Lima, Peru, 23–25 September 2024.

Abstract: Oil-in-water (O/W) nanoemulsions offer significant potential for protecting and delivering sensitive ingredients such as chia seed oil, which is rich in ω-3 fatty acids (approximately 64% α-linolenic acid, ALA). This research work aimed to study the in vitro fat digestibility of chia O/W nanoemulsions (Cas1000) with 10% (*w/w*) of chia oil and 2% (*w/w*) of sodium caseinate prepared by microfluidization (1000 bar, 3 passes) and characterized through their droplet size, superficial droplet charge, and global stability. In terms of the in vitro fat digestibility, three different matrices were studied: a water solution of sodium caseinate, a chia O/W nanoemulsion, and a bulk chia oil. The particle size distribution, mean diameter, and microstructure were evaluated after in vitro stomach and small intestine simulation according to the INFOGEST method. Free fatty acids (% FFA) produced during lipolysis were quantified at the end of digestion through their neutralization by acid-base volumetric assay. The droplet size of the Cas1000 had slight changes during the gastric phase while a significant variation of this parameter was observed at the end of the intestinal phase. A higher %FFA was obtained in Cas1000 compared to bulk chia oil with values of 58.26 and 38.13%, respectively. The ALA content in the lipid phase was quantified at the end of the gastrointestinal digestion process. The results indicated no significant changes compared to the initial oil, suggesting no losses of active compounds during digestion.

Keywords: free fatty acids; INFOGEST; lipolysis; microfluidization; omega-3 fatty acids

1. Introduction

The ω-3 and ω-6 fatty acids are essential for human health, but modern diets often have an imbalance, with higher omega-6 intake, such as linoleic acid (LA), compared to ω-3 like α-linolenic acid (ALA). This imbalance can lead to negative health effects, making it important to improve the ω-3/ω-6 ratio in the diet. Chia (*Salvia hispanica* L.) is a rich source of ALA, contributing about 64% of the total fatty acids in its oil, which helps increase ω-3 intake and rebalance this ratio, making it a valuable functional ingredient for better health outcomes [1,2].

ALA is highly susceptible to oxidation, which comprises its stability during food storage and processing. Additionally, its bioavailability in the body is limited, as it can be degraded before being effectively absorbed and utilized. These challenges underscore the need for efficient delivery systems, such as nanoemulsions, that protect ALA from oxidation and enhance its absorption in the body, ensuring that consumers can fully benefit from its health-promoting properties [1].

Emulsions can be tailored with various compositions and techniques to achieve different physicochemical and functional properties. Nanoemulsions, with droplet sizes of 20 to 200 nm, provide enhanced stability and bioavailability of lipophilic compounds, reducing gravitational separation and droplet aggregation compared to conventional emulsions, thus extending food shelf life [3].

In vitro digestibility is a key tool for assessing the breakdown and absorption of essential fatty acids, like ALA, in the gastrointestinal tract. This method simulates nutrient release and availability under controlled conditions, providing valuable insights into digestion and utilization efficiency [4].

This study investigated the in vitro digestibility of chia oil-in-water nanoemulsions, formulated with sodium caseinate as the emulsifier and microfluidization for the emulsification process.

2. Materials and Methods

2.1. Materials

Chia oil (CO) was purchased from Solazteca SDA S.A. (Buenos Aires, Argentina). Sodium caseinate (Cas), pepsin, pancreatin, and bovine bile were obtained from Sigma-Aldrich (Steinheim, Germany), and ALA and LA methyl ester standards were from Sigma-Aldrich (St. Louis, MO, USA). All other reagents were analytical grade.

2.2. Emulsion Preparation

O/W nanoemulsions (Cas1000) were prepared with 10% *w/w* of chia oil and 2% *w/w* of sodium caseinate. Pre-emulsions were formed by homogenizing CO and Cas solution with an Ultraturrax T-25 (Janke and Kunkel GmbH, Staufen, Germany) at 10,000 rpm for 2 min, then further processed using a microfluidizer (LM10, Microfluidics, Westwood, LA, USA) at 1000 bar for 3 cycles. Nanoemulsions were treated with 12 ppm nisin and 1000 ppm of potassium sorbate to inhibit microbial growth.

2.3. Emulsion Characterization

2.3.1. Particle Size and ζ-Potential

Particle size was measured in triplicate using static light scattering with a Malvern Mastersizer 2000E (Malvern Instruments Ltd., Worcestershire, UK) according to Julio et al. [3]. The ζ-potential was determined using a Zeta Potential Analyzer (Brookhaven 90 Plus/Bi-MAS, Holtsville, NY, USA) [3].

2.3.2. Physical Stability

The physical stability of emulsions was assessed with a Vertical Scan Analyzer (Quick Scan, Coulter Corp., Miami, FL, USA) during 50 d [3].

2.3.3. Microscopic Analysis

The microstructure was analyzed with a Leica DMLB optical microscope (Leica Microscopy Systems Ltd., Heerbrugg, Switzerland), using 0.2% Nile red fluorescent dye to stain the oil phase. Micrographs were captured at 63× magnification at room temperature (25 ± 1 °C).

2.4. In Vitro Digestion

In vitro digestion of chia O/W nanoemulsion (Cas1000) was performed using the static INFOGEST protocol [5], with continuous phase (Cas) and bulk chia oil (CO) as controls. Simulated gastric fluids included pepsin (2000 U/mL), and intestinal fluids contained pancreatin (100 U/mL of trypsin) and bovine bile (10 mmol/mL).

2.4.1. Particle Size and Microscopic Analysis

The size and microstructure of the particles in each system subjected to in vitro digestion at the end of the gastric and intestinal phases were evaluated as detailed in Section 2.3.

2.4.2. Extent of Lipolysis

The degree of lipolysis was determined by measuring the amount of free fatty acids (FFA) produced in CO and Cas1000 immediately after intestinal digestion according to Pinheiro et al. [6].

2.4.3. Fatty Acid Composition

The relative percentages of LA and ALA in chia oil and Cas1000 oil before and after in vitro digestion were determined by gas chromatography using an Agilent Technologies 7890 A gas chromatograph (Santa Clara, CA, USA) with a flame ionization detector and a DB-23 column. For oil extraction, 25 mL of a 1:3 isopropanol/isooctane mixture was added to 40 mL of the emulsion or digestate, vortexed, and centrifuged at 3000 rpm for 2 min. The organic phase was collected, and the solvent was evaporated using a Büchi B-480 rotary evaporator (Büchi, Switzerland) at 100 bar and 50 °C.

2.5. Statistical Analysis

The results were analyzed using ANOVA ($p \leq 0.05$) with Statgraphics Centurion XV.II software (StatPoint Technologies, Warrenton, VA, USA). Multiple comparisons were conducted with the Tukey test ($p \leq 0.05$) at a 95% confidence level.

3. Results and Discussion

3.1. Emulsion Characterization

The particle size distribution (PSD) of Cas1000 showed a bimodal pattern (Figure 1) with a narrow peak around 0.134 µm and a secondary peak at 0.63 µm. $D_{4.3}$ value was 0.15 ± 0.1 µm. At pH~6.5, the droplets had a negative surface charge of -41 ± 1 mV. Cas1000 remained physically stable, maintaining its BS profiles without significant changes over 50 d. These results indicate that the microfluidization process and emulsifier were effective in forming chia O/W nanoemulsions.

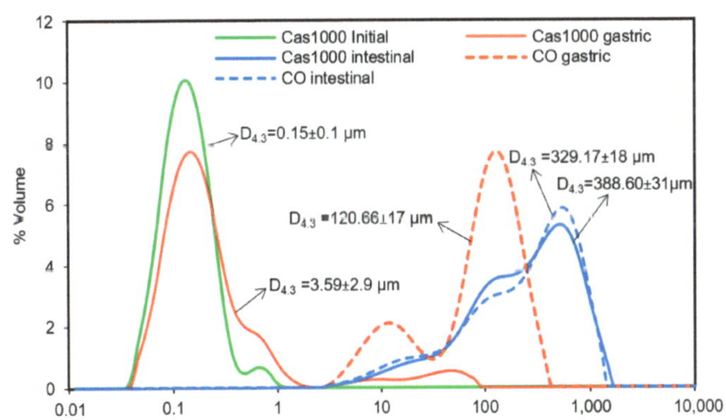

Figure 1. The particle size of Cas1000 (solid line) and CO (dotted line) at the initial time (—), after gastric (—), and intestinal (—) stages of the in vitro digestion assay. CO: chia oil; Cas1000: chia O/W nanoemulsions.

3.2. In Vitro Digestion

3.2.1. Particle Size and Microscopic Analysis

The size of oil droplets can indicate the efficiency and rate of lipolysis [7]. Figure 1 shows the PSD of Cas1000 before digestion and of both CO and Cas1000 at the end of the gastric and intestinal phases. The PSD of Cas1000 after the gastric phase was similar to the initial system (Cas1000 Initial). However, after the intestinal stage, the PSD of Cas1000 changed significantly, displaying a broad range of particle sizes and shifting towards larger sizes. Regarding CO, it exhibited bi and trimodal PSD after the gastric and intestinal phases, respectively. By the end of digestion, CO and Cas1000 had similar PSD.

The mean particle size of Cas1000, expressed as $D_{4.3}$, remained similar after exposure to the simulated gastric conditions ($p > 0.05$), indicating the stability of this system. However, this diameter significantly increased ($p \leq 0.05$) after intestinal digestion. On the other hand, CO had a significantly larger $D_{4.3}$ ($p \leq 0.05$) than Cas1000 at the end of the gastric phase. By the end of the intestinal phase, there were no significant differences ($p > 0.05$) between the two systems for this parameter (Figure 1).

Confocal microscopy images of the CO and Cas1000 after both gastric and intestinal digestion stages are shown in Figure 2. While no significant changes in particle size were observed between the initial Cas1000 and after gastric digestion using the light scattering method, the microscopy images revealed some degree of flocculation and early signs of coalescence (Figure 2). In the gastric phase, the high ionic strength and acidification of the medium (pH 3.0), passing through the isoelectric point of the proteins (~4.6), may be responsible for these changes [8]. Under these conditions, the protein layer loses sufficient repulsive forces, leading to droplet aggregation. Furthermore, confocal microscopy showed that most lipids were digested by the end of the intestinal phase since few droplets remained after this stage. Consistent with the particle size results, aggregates of various sizes observed at the intestinal stage suggest the presence of micelles, liposomes, insoluble calcium soaps, and protein complexes.

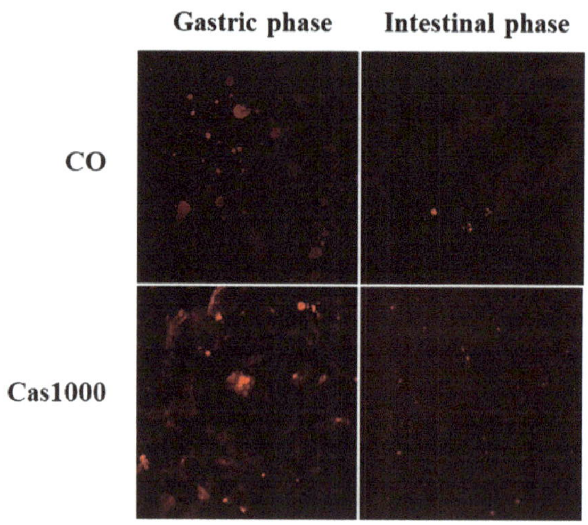

Figure 2. Confocal images (63×) of CO and Cas1000 after in vitro digestion. CO: chia oil; Cas1000: chia O/W nanoemulsions.

3.2.2. Extent of Lipolysis and Fatty Acid Composition

Lipid digestion begins in the gastric phase, and lipolysis in the intestinal phase occurs due to enzymatic action. The extent of lipolysis was assessed by measuring free fatty acids (FFA) produced during intestinal digestion. Significant differences were observed ($p \leq 0.05$), with a mean of 38.13% for CO and 58.26% for CAS1000. The higher FFA from Cas1000 suggests that emulsifying the oil droplets with protein enhances their resistance to enzymatic and acidic degradation in the gastric phase. This allows more oil to reach the intestinal phase, where lipolysis results in increased production of low molecular weight fatty acids. This behavior is consistent with that reported by Timilsena et al. [7].

Regarding fatty acid composition, CO and Cas1000 showed significant differences ($p < 0.05$) from the initial chia oil in terms of LA after the intestinal phase (Table 1). For ALA, Cas1000 did not differ significantly ($p > 0.05$) from CO, but both systems exhibited significant differences compared to the initial chia oil. These findings suggest that the digestion process impacts the fatty acid profile of the oil, likely due to changes in the emulsion stability and the extent of lipolysis.

Table 1. Fatty acid composition of initial chia oil and post-digestion OC and Cas1000.

Fatty Acid	Initial Chia Oil	Intestinal Phase	
		CO	Cas1000
Linoleic acid (%)	18.30 ± 0.50 [a]	20.88 ± 0.08 [b]	21.67 ± 0.01 [c]
α-linolenic acid (%)	67.00 ± 1.30 [b]	58.28 ± 0.09 [a]	58.48 ± 0.30 [a]

Average values ± standard deviations (n = 2). Different letters in the same row indicate significant differences ($p \leq 0.05$) between systems according to the Tukey test. CO: chia oil; Cas1000: chia O/W nanoemulsions.

4. Conclusions

Cas1000 nanoemulsions maintained high stability and consistent particle size distribution over 50 days. During digestion, Cas1000 was stable in the gastric phase but showed increased particle sizes in the intestinal phase, while CO displayed more significant changes. Cas1000 also had higher lipolysis than CO, indicating better resistance to gastric degradation and improved availability for intestinal digestion. Both systems showed changes in fatty acid composition from the initial chia oil, highlighting the impact of digestion on oil stability and bioavailability. These results underscore Cas1000's potential to enhance oil stability and bioavailability, supporting its use as a functional ingredient. Given these characteristics, Cas1000 nanoemulsions could be effectively applied in developing nutraceuticals, functional foods, and supplements where improved delivery and bioavailability of omega-3 fatty acids are crucial. Additionally, they may be valuable in formulating specialized dietary products aimed at optimizing lipid intake and digestion.

Author Contributions: Conceptualization, Methodology: L.J., G.Q.-G., E.G. and V.I.; Investigation, Formal analysis: L.J., G.Q.-G. and E.G.; Writing—review & editing, Supervision, Project administration: L.J. and V.I.; Funding acquisition: V.I. All authors have read and agreed to the published version of the manuscript.

Funding: This work was supported by Universidad de Lima—Peru, ANPCyT (PICT 2020-1274), CONICET (PIP 2007), and Universidad Nacional de La Plata (UNLP) (Project X-907), Argentina.

Institutional Review Board Statement: Not applicable.

Informed Consent Statement: Not applicable.

Data Availability Statement: Data are contained within the article.

Acknowledgments: Authors wish to thank Mariela Fernandez (CETMIC) and Mariana Pennisi (CIDCA) for their technical assistance.

Conflicts of Interest: The authors declare no conflict of interest.

References

1. Shahidi, F.; Ambigaipalan, P. Omega-3 polyunsaturated fatty acids and their health benefits. *Annu. Rev. Food Sci. Technol.* **2018**, *9*, 345–381. [CrossRef] [PubMed]
2. Ixtaina, V.Y.; Martínez, M.L.; Spotorno, V.; Mateo, C.M.; Maestri, D.M.; Diehl, B.W.; Nolasco, S.M.; Tomás, M.C. Characterization of chia seed oils obtained by pressing and solvent extraction. *J. Food Compos. Anal.* **2011**, *24*, 166–174. [CrossRef]
3. Julio, L.M.; Copado, C.N.; Diehl, B.W.; Tomás, M.C.; Ixtaina, V.Y. Development of chia oil-in-water nanoemulsions using different homogenization technologies and the layer-by-layer technique. *Explor. Foods Foodomics* **2024**, *2*, 107–124. [CrossRef]
4. Ruan, R.; Wang, J. In vitro digestion models for the study of dietary lipids: A review. *J. Food Sci.* **2020**, *85*, 3555–3567. [CrossRef]
5. Brodkorb, A.; Egger, L.; Alminger, M.; Alvito, P.; Assunção RBallance, S.; Bohn, T.; Bourlieu-Lacanal, C.; Boutrou, R.; Carrière, F.; Clemente, A.; et al. INFOGEST static in vitro simulation of gastrointestinal food digestion. *Nat. Prot.* **2019**, *14*, 991–1014. [CrossRef] [PubMed]
6. Pinheiro, A.C.; Lad, M.; Silva, H.D.; Coimbra, M.A.; Boland, M.; Vicente, A. Unravelling the behaviour of curcumin nanoemulsions during in vitro digestion: Effect of the surface charge. *Soft Matter* **2013**, *9*, 3147–3154. [CrossRef]

7. Timilsena, Y.P.; Adhikari, R.; Barrow, C.J.; Adhikari, B. Digestion behaviour of chia seed oil encapsulated in chia seed protein-gum complex coacervates. *Food Hydrocoll.* **2017**, *66*, 71–81. [CrossRef]
8. Okuro, P.K.; Viau, M.; Marze, S.; Laurent, S.; Cunha, R.L.; Berton-Carabin, C.; Meynier, A. In vitro digestion of high-lipid emulsions: Towards a critical interpretation of lipolysis. *Food Funct.* **2023**, *14*, 10868–10881. [CrossRef] [PubMed]

biology and life sciences forum

MDPI

Proceeding Paper

Production of Kefir Powdered Milk Beverage Based on Probiotic Bacteria Enriched with Lupin, Kiwicha, and Quinoa †

Nisde Nayeli Trujillo *, Nancy Ascencion Chasquibol and Silvia Patricia Ponce *

Industrial Engineering Career, Faculty of Engineering, Universidad de Lima, Av. Javier Prado Este 4600-15023, Lima 15023, Peru; nchasquibol@ulima.edu.pe

* Correspondence: 20184550@aloe.ulima.edu.pe (N.N.T.); sponce@ulima.edu.pe (S.P.P.)

† Presented at the VI International Congress la ValSe-Food, Lima, Peru, 23–25 September 2024.

Abstract: The production of functional foods has aroused growing interest due to its proven health benefits and potential to improve quality of life. One of the products that have gained importance due to its practicality is enriched beverages. Kefir, a fermented beverage traditionally produced from cow's milk, is cultivated using kefir grains containing a symbiotic culture of bacteria and yeast, which has great nutritional power and benefits the microbiota. In this research, an enriched powdered milk beverage with a high protein content is prepared due to the incorporation of lupine (*Lupinus mutabilis*), kiwicha (*Amaranthus caudatus*), and quinoa (*Chenopodium quinoa*). The beverage prepared shows a 48% increase in protein content compared to commercialized kefir. After the beverage was obtained, it was freeze-dried to preserve its nutritional and functional properties. The resulting beverage, kefir milk powder, enriched with native Andean grains such as lupin, kiwicha, and quinoa, presents a sustainable and nutrient-rich option that contributes to dietary diversification.

Keywords: fermentation; kefir; kiwicha; probiotic bacteria; proteins; quinoa; lupin

1. Introduction

The development of novel functional and probiotic beverages is a growing trend in the food industry, driven by the increasing demand for healthier products [1]. One of the most widely consumed foods globally is dairy products, which are beneficial for health owing to their calcium content; kefir, a fermented dairy drink, is particularly beneficial for gut health [2]. Its nutritional profile can be further enhanced by adding Andean grains such as lupin, kiwicha, and quinoa, which are well-known for their high protein content [3]. Previous research has employed lupin to enhance the nutritional value of fresh cow's milk, improving its protein content by 31.03% [4]. Mendoza [5] prepared a functional beverage based on quinoa and pineapple, rich in minerals and proteins, which exhibited a 6.4-fold increase in protein content compared to a conventional pineapple beverage. Quinoa (*Chenopodium quinoa*), kiwicha (*Amaranthus caudatus*), and lupin (*Lupinus mutabilis*) are known for their high nutritional value, including lipids, proteins, dietary fiber, and a wide range of minerals. Therefore, the combination of kefir and these legumes can result in a highly nutritious product. This research focuses on the production of a functional probiotic beverage based on kefir, a promising source of proteins and nutrients, using quinoa, kiwicha, and lupin.

2. Materials and Methods

2.1. Preparation and Fermentation of Kefir

One liter of whole cow's milk was preheated to 25 °C and transferred to a glass container. Twenty grams of kefir grains, a consortium of acetic and lactic acid bacteria (LAB) and yeasts encompassing 20 different strains, were added. The flask was covered with sterile muslin cloth and incubated at 25 °C in the dark. After 24 h, the liquid was separated from the kefir grains using a sterile plastic sieve. Kefir grains that passed through the sieve were washed with sterile water and returned to fresh whole milk. All ingredients were purchased from a local market in Lima, Peru.

2.2. Enrichment with Lupin, Kiwicha and Quinoa

Lupin grains were soaked for 12 h to remove bitterness. Subsequently, 150 g of lupin grains, 30 g of kiwicha flour, and 70 g of quinoa flour were added to 700 mL of kefir milk. The mixture was blended and pasteurized at 90 °C for 15 min. All ingredients were purchased from a market in Lima, Peru.

2.3. Freeze-Drying of the Final Product

The sample was subjected to a vacuum sublimation process at −50 °C for 24 h in a freeze-dryer (Martin Christ Alpha 1–2, Zirbus VaCo 5) to obtain the enriched kefir drink powder. The powder obtained (EKM) was stored in sealed polyethylene bags under vacuum conditions for quality analysis.

2.4. Proximal Composition

Proximate composition was determined according to official methods. Moisture content was determined at 110 °C to constant weight. Ash content was determined by incineration at 550 °C for 72 h in a muffle furnace. Total protein content was determined using a Kjeldahl analyzer (UDK 139, VELP, Usmate Velate, Monza, Italy). Fat content was determined using n-hexane extraction for nine hours. All measurements were performed in triplicate [6].

2.5. Determination of Mineral Content

The concentrations of Ca, Cu, Fe, K, Mg, Na, P, and Zn in the powdered beverage were determined by atomic absorption spectrophotometry (Nexion 350×, Perkin Elmer, MA, USA) [6].

2.6. Physical-Chemical Properties

Apparent and compacted densities, Hausner ratio, and solubility were determined according to the methodologies outlined by Chasquibol et al. [6]. Hygroscopicity was determined according to the methodology described by Nowshina et al. [7]. A Brookfield rotational rheometer (model DV-II Pro LV; Brookfield Engineering Laboratories, Middleboro, MA, USA) was used for viscosity analysis. The data obtained were processed using the Rheocalc application software (V3.2 Build 47-0; Brookfield Engineering Laboratories, Middleboro, MA, USA) according to the methodology described by Kok-Tas et al. [8].

2.7. Color Measurements

Color measurements were performed using a Color Tec-PCM colorimeter (Model D25-PC2, Cole Palmer, Vernon Hills, Chicago, IL, USA) according to the methodology described by Nowshina et al. [7].

2.8. Amino Acid Determination

The amino acid profile was quantified by HPLC (ARC, Waters, Milford, CT, USA) following the methodology outlined by Chasquibol et al. [6]. All measurements were performed in triplicate.

2.9. Sensory Analysis

Fifteen semi-trained panelists evaluated the sensory analysis (color, flavor, aroma, texture, and overall acceptability) of the powder following the methodology presented by Nowshina et al. [7].

2.10. Statistical Analysis

Results were expressed as mean ± standard deviation. All measurements were performed in duplicate or triplicate. Analysis of variance (ANOVA) was used for the mixture design employed in this study and for response optimization of the protein variable. Data were analyzed at a 95% significance level using Minitab 19.0 statistical software (Minitab Inc., State College, Palo Alto, PA, USA).

3. Results and Discussion

3.1. Statistical Analysis

A simplex centroid mixture design was used to optimize the protein content in a formulation composed of kefir milk enriched with lupin, kiwicha, and quinoa, as depicted in Table 1. The design consisted of five experimental runs, varying the levels of kefir milk (700–800 mL) and lupin (50–150 g). The results indicated that the maximum protein content was achieved with a mixture containing 700 mL of kefir milk and 150 g of lupin, as depicted in Figure 1. The quantities of kiwicha and quinoa were held constant throughout the experiment.

Table 1. Statistical Analysis.

RunOrder	PtType	Blocks	Kefir Milk	Lupin	Proteins Content
1	−1	1	0.725	0.125	32.27
2	1	1	0.700	0.150	37.92
3	−1	1	0.775	0.075	19.29
4	0	1	0.750	0.100	23.86
5	1	1	0.800	0.50	14.70

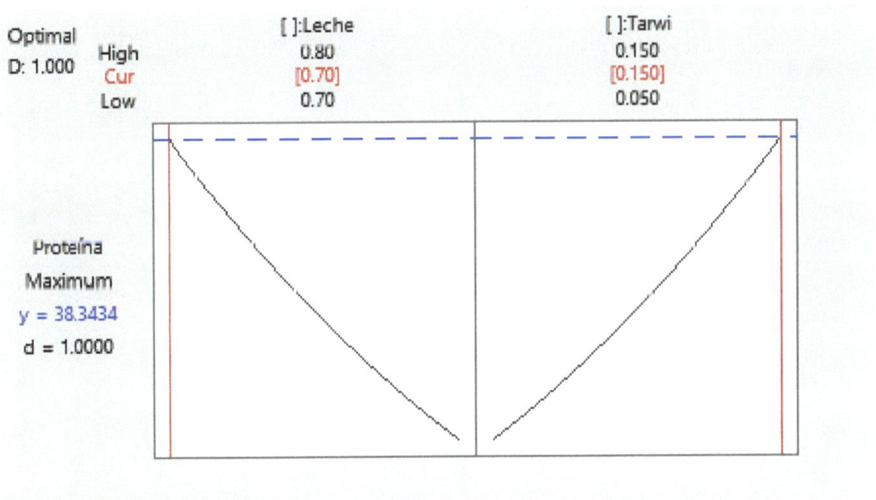

Figure 1. Predicted responses plot.

3.2. Proximal Composition and Mineral Content, Recommended Dietary Intake (RDI), and Upper Tolerance Limit (UTL)

The Andean grains proved to be highly effective in augmenting the protein content of the product, as evidenced by the data presented in Table 2.

Table 2. Proximal composition, mineral content, recommended dietary intake (RDI), and upper tolerance limit (UTL).

Proximal Composition (%)	EKM (g/100 g)	Minerals	EKM (mg/kg)	RDI * (mg/day)	UTL (mg/day)
Moisture	0.34 ± 0.002	Calcium	2505.50 ± 4.02	1000	2500
Ash	3.59 ± 0.02	Copper	4.20 ± 0.02	0.9	8
Protein	37.92 ± 0.02	Iron	42.30 ± 0.10	W 18, M 8	45
Fat	18.82 ± 0.02	Potassium	5877.11 ± 7.04	W 2600, M 3400	3400
		Magnesium	1375.96 ± 3.07	W 315, M 410	350
		Sodium	1644.69 ± 2.01	2000	2300
		Phosphorus	4799.81 ± 3.02	700	4000
		Zinc	31.53 ± 0.07	W 8, M 11	40

Values are mean ± SD; n = 3. * For women (W) and men (M), aged 19 to 50 years.

The protein content increased from 14.70 g to 37.92 g. Furthermore, the moisture content was measured to be 0.34 ± 0.002 g per 100 g. The freeze-drying method can eliminate all the liquid content of the sample and the storage of the product has to be in sealed polyethylene bags under vacuum conditions to prevent an increase in moisture, bacterial activity, and the degradation of the product. The determination of the ash content of the EKM showed a 3.59 ± 0.02 g/100 g product.

EKM contains a wide range of minerals, as indicated in Table 2; if it is compared with the commended daily intake for each mineral (National Institute of Health, 2023) [6], the consumption of only 20 g daily of EKM would cover 5.01% of the daily requirement of calcium, 12.29% of iron, 4.52% of potassium, 8.73% of magnesium, and 13.71% of phosphorus. Therefore, EKM offers a nutritious option full of various minerals that contribute to a balanced diet.

3.3. Physical-Chemical Analyses

Table 3 shows physical-chemical analyses of the product (EMK). The solubility value of EKM obtained was 26.96% ± 0.02 due to the higher fiber content of lupin, kiwicha, and quinoa. The value for hygroscopicity was 0.03 ± 0.02, indicating that the EKM sample has a lower capacity to absorb moisture from the environment, implying reduced absorption of vapor or liquid. This low moisture content in powdered milk inhibits the growth and reproduction of microorganisms. Since hygroscopic ingredients are sensitive to moisture, and in powdered milk, it can lead to deterioration, such as softening, color changes, and disintegration [2]. Therefore, the water activity in EKM was e 0.86 ± 0.01, which agrees with the aforementioned findings. By maintaining low moisture content, the proliferation of bacteria, molds, and yeasts is effectively controlled. The EKM exhibited an apparent density of 6.9 ± 0.01 g/mL and a compacted density of 4.2 ± 0.03 g/mL, with a Hausner ratio 0.6. It also possessed a viscosity of 20.06 mPa·s, indicating optimal fluidity. These values suggest that the powdered beverage has good flowability.

Table 3. Physical-chemical properties of the product (EMK).

Analysis	EKM
Solubility %	26.96% ± 0.002
Hygroscopicity (g)	0.03 ± 0.002
Water activity	0.86 ± 0.001
Bulk density (g/cm^3)	6.9 ± 0.001
Tapped density (g/cm^3)	4.2 ± 0.003
L*	78.1 ± 1.2
a*	−1.59 ± 0.02
b*	4.89 ± 0.02

Values are mean ± SD; n = 3.

For the color analysis of EKM, the L*, a*, and b* values of the samples were analyzed after preparation according to the formula. The EKM shows a higher value of luminosity (L *); this is because it is made from cow's milk, and it retains a color close to chalk white, which is quite similar to quinoa and kiwicha flour; the lupin gives it a yellowish cream appearance; that gives color to EKM a color is obtained in the range of white, lead and cream, a color close to ivory, which matched to milk measurements. The a* represents the redness or greenness of the products; in EKM, a negative value was shown, which represents the greenness. The b* value represents yellowness values or a blue indicator; in EKM, a positive value is shown, which represents yellowness in the sample. The low values of greenness and yellowness are due to the original color of the lupin in EKM [7].

3.4. Amino Acids in the Product

Table 4 presents the results of the amino acid profile of the product, a sample of kefir milk powder enriched with lupin, kiwicha, and quinoa (EKM). The protein properties of lupin, kiwicha, and quinoa have benefited the EKM by allowing them to retain better concentrations of essential amino acids. Essential amino acids cannot be synthesized within the body and, therefore, must be provided through the diet [9]. The predominant amino acids in EKM are glutamic acid (85.5 ± 0.26 mg/g of protein), arginine (46.14 ± 0.14 mg/g of protein), and leucine (39.63 ± 0.22 mg/g of protein), which is, according to informed by Akal [9], aspartic acid (37.61 ± 0.04 mg/g protein), NH 2 (27.93 ± 0.30 mg/g protein), and proline (25.24 ± 0.43 mg/g protein); the limiting amino acids are methionine 7.62 ± 0.22 mg/g protein and tryptophan 4.05 ± 0.55 mg/g protein. The amino acid composition of EKM showed a balanced profile in accordance with FAO/WHO recommendations and demonstrated that dairy drinks formulated with lupin, kiwicha, and quinoa are rich in amino acids, except for tryptophan [7].

Table 4. Amino acid profile of the product (mg/g protein).

Aminoacid	EKM
Aspartic Acid	37.61 ± 0.04
Glutamic Acid	85.80 ± 0.26
Serine	21.55 ± 0.08
Histidine	8.31 ± 0.22
Glycine	18.61 ± 0.40
Threonine	16.09 ± 0.47
Arginine	46.14 ± 0.14
Alanine	13.93 ± 0.06
Proline	25.24 ± 0.43
Tyrosine	14.02 ± 0.35
NH_3	27.93 ± 0.30
Valine	21.55 ± 0.35
Methionine	7.62 ± 0.22
Cystine	1.58 ± 0.01
Isoleucine	17.28 ± 0.45
Leucine	39.63 ± 0.22
Phenylalanine	19.09 ± 0.12
Lysine	21.10 ± 0.36
Tryptophan	4.05 ± 0.55

Values are mean ± SD; n = 3.

3.5. Sensory Evaluation

Table 5 shows the results of 15 panelist's evaluation of the organoleptic characteristics for EKM.

EKM obtained a score of 7.1 for color, indicating a similar color to commercial milk. This is in contrast to the color indicator reported by Nowshina et al. [7], which ranged from 5.3 to 6.7 due to the addition of maltodextrin and sunflower oil, causing a color change from yellow to brown, and was not well-accepted by their panelists. Regarding flavor,

EKM scored 6.7, as it maintained a neutral taste (similar to almond milk, according to four panelists). EKM received a score of 7.9 for aroma, characterized by a unique flavor profile typical of kefir milk, with a predominant aroma of quinoa and kiwicha, which were highly appreciated by the panelists. The texture was rated 7.3, as it was described as a fluid and easy-to-drink beverage. EKM achieved an overall acceptability score of 7.2 (on a 1–9-point rating scale), aligning with the panelists' perceptions of color, flavor, aroma, and texture.

Table 5. Average score of overall acceptability of the product (EKM).

Formulation	EKM
Color	7.1 ± 0.02
Flavor	6.7 ± 0.02
Arome	7.9 ± 0.02
Texture	7.3 ± 0.02
Overall Acceptability	7.2 ± 0.03

4. Conclusions

The formulation developed of kefir milk powder enriched with lupin, kiwicha, and quinoa enabled obtaining a beverage with high protein content. The addition of these Andean grains enhanced the product's mineral content, particularly calcium, which is essential in dairy beverages. Moreover, it provided a significant amount of essential amino acids. The powder exhibited acceptable solubility with low hygroscopicity, moisture content, and water activity, thereby preventing microbial growth and improving product quality and shelf life. Sensory evaluation revealed acceptable characteristics in terms of color, flavor, texture, and aroma, indicating a generally positive consumer acceptance. In conclusion, kefir milk powder enriched with lupin, kiwicha, and quinoa represents a nutritious option, providing a significant source of protein and probiotic benefits to improve gut health and overall nutrition.

Author Contributions: All authors have contributed equally to this manuscript. Conceptualization, N.N.T., N.A.C. and S.P.P.; Methodology, N.N.T., N.A.C. and S.P.P.; Investigation and Data analysis, N.N.T., N.A.C. and S.P.P.; Writing—original draft preparation, N.N.T., N.A.C. and S.P.P.; Writing—review and editing, N.N.T., N.A.C. and S.P.P. All authors have read and agreed to the published version of the manuscript.

Funding: This research received no external funding.

Institutional Review Board Statement: Not applicable.

Informed Consent Statement: Not applicable.

Data Availability Statement: Data is contained within the manuscript.

Acknowledgments: This work was supported by Laboratorio de Alimentos Funcionales of Carrera de Ingeniería Industrial, Universidad de Lima—Peru.

Conflicts of Interest: The authors declare no conflict of interest.

References

1. Vinícius, K.; Cortina, V.; Camelo, C.; Amaral, L.; Da Silva, A.; Verruck, S.; Quintaño, L. Powdered water kefir: Effect of spray drying and lyophilization on physical, physicochemical, and microbiological properties. *Food Chem. Adv.* **2024**, *5*, 100759. [CrossRef]
2. M'hir, S.; Ziadi, M.; Mejri, A.; Ayed, L. Mixture of whey-milk and palm sap for novel kefir beverage using simplex-centroid mixture design. *Kuwait J. Sci.* **2023**, *50*, 690–696. [CrossRef]
3. Jamanca, N.; Ocrospoma, R.; Eguilas, Y.; Padilla, R.; Silva, R. Food grain quality: Analysis of physical, biometric, and colorimetric properties to promote consumption. *Heliyon* **2024**, *10*, e29234. [CrossRef] [PubMed]
4. Márquez, L.; Pretell, C.; Hayayumi, M. Functional beverage design based on fresh milk, tarwi (*Lupinus mutabilis*) beverage and oatmeal (*Avena sativa*). *Scielo Anal.* **2021**, *48*, 490–499. [CrossRef]

5. Mendoza, V. Productos Funcionales: Oportunidades para el Peru. Ministerio de Desarrollo Agrario y Riego. 2021. Available online: https://repositorio.sierraexportadora.gob.pe/bitstream/handle/SSE/263/Funcionales%20-%20Mayo.pdf?sequence=1 (accessed on 28 May 2024).
6. Chasquibol, N.; Alarcón, R.; Sotelo, A. Development of Powdered Beverage with Cushuro (*Nostoc commune*) Concentrated Protein and Quinoa (*Chenopodium quinoa*). *Biol. Life Sci. Forum* **2023**, *25*, 2. [CrossRef]
7. Nowshina, H.; Farzana, A.; Islam, A.; Rahim, A.; Begum, A.; Gulzarul, M.; Rahman, A. Nutritional, physical and sensorial properties of food-grade emulsion powder enriched with soy milk and whey protein permeates. *Food Humanit.* **2023**, *1*, 769–776. [CrossRef]
8. Kok-Tas, T.; Sedym, A.; Ozer, B.; Zeynep, B. Effects of different fermentation parameters on quality characteristics of kefir. *Dairy Sci.* **2013**, *96*, 780–789. [CrossRef] [PubMed]
9. Akal, C. Benefits of Whey Proteins on Human Health. In *Dairy in Human Health and Disease Across the Lifespan*; Academic Press: Cambridge, MA, USA, 2017; pp. 363–372. [CrossRef]

biology and life sciences forum

MDPI

Proceeding Paper

Study of the Release Kinetics of Capsaicin Extracted from Charapita Chili (*Capsicum frutescens*) from an O/W Emulsion Made with Sacha Inchi Oil (*Plukenetia volubilis*) and Encapsulated in Calcium Alginate Beads †

Nelson Tapia *, Winnie Perez, Liset Navarro, Holger Maldonado, Ale Borja and Gerardo De Lama

Grupo de Investigación GELPOL, Facultad de Química e Ingeniería Químca, Universidad Nacional Mayor de San Marcos, Av. Germán Amezaga 375, Lima 150101, Peru; winnieperezv@gmail.com (W.P.); liset.navarro@unmsm.edu.pe (L.N.); hmaldonadog@unmsm.edu.pe (H.M.); naleb@unmsm.edu.pe (A.B.); gdelamac@unmsm.edu.pe (G.D.L.)

* Correspondence: ntapiah@unmsm.edu.pe

† Presented at the VI International Congress la ValSe-Food, Lima, Peru, 23–25 September 2024.

Abstract: Capsaicin has multiple applications such as analgesic for muscle pain, anti-inflammatory, anticancer, antidepressant, and others. It is a lipophilic compound that produces irritation, therefore, for its application, it is necessary to encapsulate it in emulsions or biopolymers. The objective of this work was to prepare a direct O/W emulsion containing capsaicin extracted from Charapita chili (*Capsicum frutescens*) with Sacha Inchi (*Plukenetia volubilis*) oil as the oil phase and encapsulated in calcium alginate beads, intending to increase the useful life of the capsaicin. Capsaicin was extracted from Charapita chili powder using ethanol, then the extract was dried for 40 min at 55 °C. To obtain the emulsion, the dry extract was dissolved in 100 mL of Sacha Inchi oil and mixed with sodium alginate by stirring at 14,000 rpm. Then, the emulsion was combined with sodium alginate by stirring at 14,000 rpm. This mixture was dripped onto 200 mL of a 0.2 M solution of calcium chloride, obtaining beads of a spherical shape. The experimental kinetic data are described with the Korsmeyer–Peppas model, where the maximum release was reached at 180 min, the value of the constant **n** was 0.7857, and the rate constant K was 2.95. As the constant $n < 0.85$, the release process was due to the diffusion and swelling of the beads. The emulsion obtained could be used to develop pharmaceutical products; moreover, the encapsulated emulsion in calcium alginate could be used in the formulation of functional foods to take advantage of the capsaicin from Charapita chili and the functional properties of the Sacha Inchi oil.

Keywords: capsaicin; Charapita chili; encapsulation; emulsion; kinetic

1. Introduction

There is a great variety of chili peppers in Peru, which have been cultivated by our ancestral cultures and have made Peruvian cuisine famous [1]. However, the scientific study of our chili peppers is still incipient. Among these chili peppers is the Charapita chili (*Capsicum frutescens*), which is characterized as a small yellow fruit that is very aromatic and spicy. It is grown in the Peruvian Amazon, in the areas of Loreto, San Martin, and Ucayali.

Currently, the extraction of capsaicin from chili fruits is carried out using Soxhlet extraction equipment, in which methanol, ethanol, acetone, etc., are used as solvents [2]. In addition, to carry out the extraction, ultrasound, microwave, and supercritical fluid techniques are used. Capsaicin molecules (8-methyl-N-vanillyl-trans-6-nanenamide) are hydrophobic vanilloide compounds that give spicy foods their pungent quality. It is a secondary metabolite found in the composition of different varieties of chili peppers with high therapeutic and nutritional potential because they contain a wide range of bioactive compounds, such as capsaicinoids, carotenoids, polyphenols, flavonoids, vitamins, and minerals [3].

Regarding therapeutic applications, Muwen et al. [4] carried out a review on capsaicin as an analgesic in chronic pain caused by osteoarthritis and rheumatoid arthritis. Capsaicin has anti-inflammatory characteristics, which are widely used in gels, creams, and ointments. As an anticancer compound, Willian et al. [5] highlighted that capsaicin exhibits potent anticancer effect on several types of human cancer, including gastric cancer, breast cancer, lung cancer, prostate cancer, etc.

The direct oral administration of capsaicin causes adverse effects in our body because it produces irritation in the oral cavity and stomach, which leads to the emergence of stomatitis and gastric ulcers [6]. In addition, capsaicin has low solubility in water, a short half-life, and cytotoxicity at high concentrations. Due to these disadvantages, administration systems are being developed that are based on different encapsulation techniques, such as emulsions, liposomes, micelles, nanoparticles, etc. The encapsulation of active ingredients in micro- and nanoemulsions are widely used to improve the oral availability of hydrophobic ingredients and are prepared by using high speed and pressure homogenizers [7].

The objective of this research was to study the encapsulation of capsaicin emulsion obtained from the Charapita chili.

2. Materials and Methods

2.1. Collection and Processing

The Charapita chili was purchased at the market of the fishing terminal Villa Maria del Triunfo in Lima, Peru. One kilogram of the fruit was cut into halves and placed in trays of a food dryer (IKE, Foshan City, Guangdong, China) for 4 h at 50 °C. The dried sample was then ground with a stainless-steel blade mill to obtain the chili powder.

2.2. Capsaicin Extraction Process

The method proposed by Hoyos [2] was carried out with some modifications. First, 6 g of the powdered material was placed in a filter paper bag, which was then introduced into the Soxhlet apparatus. For the extraction of capsaicin, 100 mL of absolute ethanol was used. The extraction of capsaicin was carried out for 6 h, maintaining the temperature of the heating blanket at 50 °C. The extract was dried in a Rotavapor equipment (D-Lab, Beijing, China) for 40 min at 55 °C. All extractions were performed in triplicate.

2.3. Emulsion Preparation Process

The emulsion preparation was performed according to the method proposed by Chan [8]. First, 0.1 g of dry extract was dissolved in 100 mL of Sacha Inchi oil, which had been bought from a local market in Lima city, Peru, by constant stirring at 500 rpm. Then, 10 mL of capsaicin solution was mixed with 40 mL of sodium alginate solution (1.6%) and the mixture was placed in homogenization equipment (VELP, Usmate Velate, Italy) at 14,000 rpm for 3 min to obtain a stable microemulsion.

2.4. Emulsion Encapsulation Process

The microemulsion was combined with 40 mL of sodium alginate by stirring at 14,000 rpm, and then this mixture, with the help of a peristaltic pump, was dripped onto a 200 mL solution of calcium chloride (0.2 M) with constant stirring at 250 rpm for 30 min to obtain beads of a spherical shape. To remove excess calcium chlorine, the beads were washed with distilled water and then dried in a food dryer (IKE, Guangdong, China) at 40 °C for 30 min, as shown in Figure 1.

Figure 1. Calcium alginate loaded with capsaicin.

2.5. Capsaicin Release Kinetics

The process began with 5 g of beads in 300 mL of absolute ethanol, and the mixture was kept under constant agitation at 300 rpm throughout the course of the kinetics. To follow the kinetics of the capsaicin release process, 5 mL aliquots were taken at different time intervals. To determine the concentration of capsaicin in each aliquot, the spectrophotometric method (Thermo Fisher Scientific, Waltham, MA, USA) was used, measuring the absorbance at the wavelength $\lambda = 281$ nm. The release kinetics of capsaicin corresponded to concentration versus time and was obtained based on the calibration curve, as shown in Figure 2.

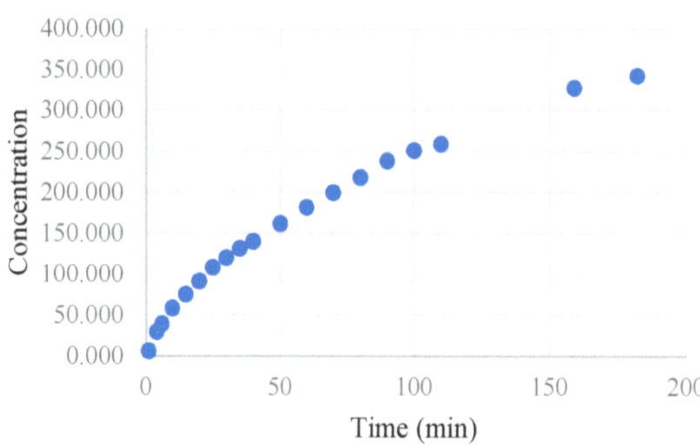

Figure 2. Capsaicin release kinetics. Concentration versus time.

3. Results and Discussion

A stable direct O/W emulsion of capsaicin was obtained, in which the emulsifying agent was the biopolymer sodium alginate. When the concentration of the biopolymer is in excess, the emulsion can be directly encapsulated through a cross-linking reaction with calcium ions, Ca^{2+}, resulting in homogeneous and resistant spherical beads. Figure 2 shows the concentration of capsaicin released as a function of time. From this result, it can be deduced that the maximum capsaicin is reached after 180 min.

Emulsions are used to transport substances with nutritive and active principles [4,5]. Chan [8] used palm oil to obtain a direct O/W emulsion and employed calcium alginate beads in its encapsulation. In this research, a stable direct O/W emulsion carrying capsaicin was obtained, and a sodium alginate biopolymer was used as an emulsifying agent instead of calcium alginate. In addition, the same method was used but with Sacha Inchi (*Plukenetia volubilis*) oil and capsaicin. The encapsulated emulsion increased the slow release of capsaicin with the objective of increasing its stability for future application in the food and pharmaceutical industry.

The release kinetics are described by the Korsmeyer–Peppas model [9], whose linear form is expressed through Equation (1), where the constant **n** indicates the release mechanism, K is the velocity constant, and the relation $C_t/C_{\inf}$ represents the fraction of capsaicin released. The results of the analysis of the experimental data based on Equation (1) using the least squares statistical method is as shown in Figure 3.

$$\ln\left(\frac{C_t}{C_{\text{inf}}}\right) = \ln K + n \cdot \ln t \tag{1}$$

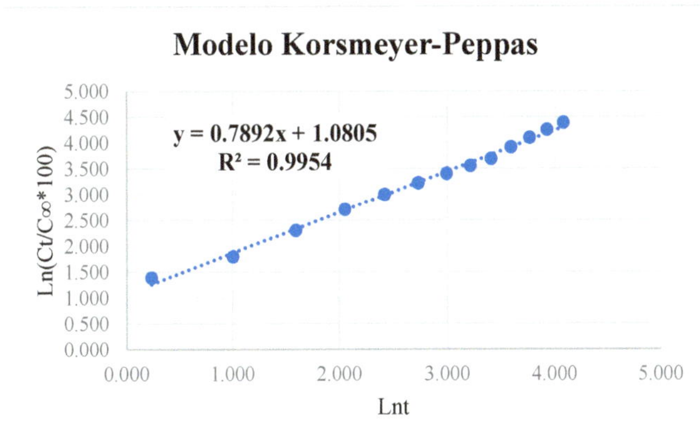

Figure 3. Korsmeyer–Peppas linear Equation (1).

According to the value obtained for the mean square deviation R^2 = 0.9954, it is deduced that the kinetic data are adequately described by the Korsmeyer–Peppas model. Furthermore, the value of n = 0.7892 ($n \leq 0.85$) indicates that the release mechanism of capsaicin from the internal part of the beads corresponds to the diffusion and swelling of the beads.

4. Conclusions

Capsaicin was extracted from Charapita chili using absolute ethanol as the solvent. A direct O/W microemulsion containing capsaicin was prepared using sodium alginate as an emulsifier agent. The emulsion was encapsulated in alginate beads due to excess sodium alginate. The release kinetics of capsaicin from the calcium alginate beads in the ethyl alcohol medium determined that the maximum release was reached at 180 min. The kinetic data were adequately described with the Korsmeyer–Peppas model, and the value of the constant n = 0.7857 indicated that the release mechanism of capsaicin from the beads corresponded to the mechanisms of diffusion and swelling of the beads. It is necessary to continue the applications of capsaicin emulsions in the pharmaceutical and food industry.

Author Contributions: All authors have contributed equally to this manuscript. Conceptualization, N.T., W.P., L.N., H.M., A.B. and G.D.L.; Methodology, N.T., W.P., L.N., H.M., A.B. and G.D.L.; Investigation and Data analysis, N.T., W.P., L.N., H.M., A.B. and G.D.L.; Writing original draft preparation, N.T., W.P., L.N., H.M., A.B. and G.D.L.; Writing-review and editing, N.T., W.P., L.N., H.M., A.B. and G.D.L. All authors have read and agreed to the published version of the manuscript.

Funding: This research was funded by Universidad Nacional Mayor de San Marcos grant number C24072301 and the APC was funded by Universidad Nacional Mayor de San Marcos.

Institutional Review Board Statement: Not applicable.

Informed Consent Statement: Not applicable.

Data Availability Statement: Data are contained within the article.

Acknowledgments: Thanks to the Universidad de Lima for the organization of the congress. This work was supported by the Vice Rector for Research and Postgraduate Studies of the Universidad Nacional Mayor de San Marcos (UNMSM), Lima-Peru by financial support of the project C24072301.

Conflicts of Interest: The authors declare no conflict of interest.

References

1. Universidad Nacional Agraria La Molina. *Ajíes Peruanos, Sazón para el Mundo*, 1st ed.; Sociedad Peruana de Gastronomía: Lima, Perú, 2009; Available online: https://www.librosperuanos.com/libros/detalle/16842/Ajies-peruanos-Sazon-para-el-mundo (accessed on 1 June 2024).
2. Hoyos Olga, L.; Franco, J.M.; Solarte, C.E.; Orozco, M.I. Extracción y cuantificación de capsaicina a partir del fruto del ají (*Capsicum* spp.). *Sci. Tech.* **2007**, *33*, 33–35. Available online: https://www.researchgate.net/publication/230724632_Extraccion_y_cuantificacion_de_capsaicaps_a_partir_del_fruto_del_aji_Capsicum_spp (accessed on 1 June 2024).
3. Liu, Z.; Cai, S.; Zhang, S.; Xiao, Y.; Devahastin, S.; Guo, C.; Wang, Y.; Wang, T.; Yi, J. A systematic review on fermented chili pepper products: Sensorial quality, health benefits, fermentation microbiomes, and metabolic pathways. *Trends Food Sci. Technol.* **2023**, *141*, 104189. [CrossRef]
4. Lu, M.; Chen, C.; Lan, Y.; Xiao, J.; Li, R.; Huang, J.; Huang, Q.; Cao, Y.; Ho, C.T. Capsaicin-the Major Bioactive Ingredient of Chili Peppers: Bio-efficacy and Delivery Systems. *Food Funct.* **2020**, *11*, 2848–2860. [CrossRef] [PubMed]
5. Rollyson, W.D.; Stover, C.A.; Brown, K.C.; Perry, H.E.; Stevenson, C.D.; McNees, C.A.; Ball, J.G.; Valentovic, M.A.; Dasgupta, P. Bioavailability of capsaicin and its implications for drug delivery. *Control. Release* **2014**, *196*, 96–105. [CrossRef]
6. Smutzer, G.; Jacob, J.C.; Tran, J.T.; Shah, D.I.; Gambhir, S.; Devassy, R.K.; Tran, E.B.; Hoang, B.T.; McCune, J.F. Detection and modulation of capsaicin perception in the human oral cavity. *Physiol. Behav.* **2018**, *194*, 120–131. [CrossRef] [PubMed]
7. Jiajia, R.; McClements, D.J. Food grade micro emulsions, nano emulsions and emulsions: Fabrication from sucrose Monopalmitato and lemon oil. *Food Hydrocoll.* **2011**, *25*, 1413–1423. [CrossRef]
8. Chan, E.S. Preparation of Ca-alginate beads containing high oil content: Influence of process variables on encapsulation efficiency and bead properties. *Carbohydr. Polym.* **2011**, *84*, 1267–1275. [CrossRef]
9. Peppas, N.A.; Sahlin, J.J. A simple equation for the description of salute release. III. Coupling of diffusion and relaxation. *Int. J. Pharm.* **1989**, *57*, 169–172. [CrossRef]

biology and life sciences forum

MDPI

Proceeding Paper

Effect of Disinfection and Drying of Wild Carob Pods (*Neltuma* sp.) on the Safety of the Carob Flour †

Karen Martínez, Natasha Sotto, Rocio Villalba, Silvia Caballero and Laura Mereles *

Dirección de Investigaciones, Facultad de Ciencias Químicas, Universidad Nacional de Asunción, San Lorenzo P.O. Box 1055, Paraguay; kmartinez@qui.una.py (K.M.); natashasotto83@gmail.com (N.S.); rvillalba@qui.una.py (R.V.); scaballero@qui.una.py (S.C.)

* Correspondence: lauramereles@qui.una.py

† Presented at the VI International Congress la ValSe-Food, Lima, Peru, 23–25 September 2024.

Abstract: In the hostile and challenging environment of the Paraguayan Chaco, the wild carob pods (*Neltuma* spp.) are a valuable vegetable resource that provides nutrition and significant economic opportunities for the local populations by means of carob flour production. However, the microbiological quality of the carob flour is limited due to the manual gathering. The main objective of this investigation is to find an efficient disinfectant and its minimum application level to obtain microbial stability in carob flour. The microbial load (total mesophilic aerobes, moulds, and yeasts, *Escherichia coli*, *Staphylococcus aureus*, and *Salmonella* spp.) of the flour obtained by disinfection with citric acid (1 and 3%) and sodium hypochlorite (1 and 3%) was compared. Drying tests were carried out at time intervals of 2, 4, 6, and 7 h on whole carob pods to obtain flour in a hot air circulating tray type dryer, and humidity was used as a response variable in a thermobalance (desirable humidity < 5%). A combined process of disinfection with 3% citric acid and hot air circulating tray-type drying for 7 h at 60 °C is proposed to obtain an innocuous carob flour of high microbiological quality.

Keywords: carob; disinfection; drying: moisture; biodiversity; *Neltuma* spp.; *Prosopis* spp.; food safety

1. Introduction

The fruits of the various species of carob tree in the Paraguayan Chaco are very important for the food and subsistence of different indigenous communities; it occupies a prominent place in the local culture and is appreciated for its significant nutritional value, especially for flour production by milling in artisanal "palo santo" mortars [1]. However, previous studies have shown that quantified levels of yeasts and mesophilic aerobes require attention to ensure the safety of carob pods that are collected from the soil manually before drying for flour production [2], which can be exposed to a variety of harmful bacteria, such as *Salmonella* spp. and pathogenic *Escherichia coli* (E. coli) [3], to fungal contamination *Aspergillus* section *Flavi* and aflatoxins and should be treated to ensure safety prior to commercialisation, as is required for other traditional crops [4].

The main objective of this work was to find an efficient disinfectant and its minimum application level to obtain microbial stability in carob flour.

2. Materials and Methods

2.1. Experiment Development

The samples of carobs collected in the Pykasu community in the Mariscal Estigarribia district, Boquerón Department of the Western Region, Paraguayan Chaco, were used. They were transported under appropriate conditions at room temperature to the Industrial Microbiology Laboratory of the Research Department of the Faculty of Chemical Sciences of the UNA. First, the washing of whole carob pods was studied under six conditions: citric acid 1% and 3%, sodium hypochlorite 0.1% and 0.3%, rinse with water, and without rinsing (control).

Each solution had a contact time of 30 min with the carob pods and a draining time of 30 min. Subsequently, all groups were treated with the same method of drying. A dryer was used in trays with forced air circulation at a constant temperature of 60 °C until reaching 5% humidity in accordance with current regulations for carob flour [5]. The dyer equipment was designed and built in the Department of Industrial Applications of the Faculty of Chemical Sciences, equipped with (a) a helical fan, (b) a table contactor, (c) an internal resistance type heater, and (d) a drying chamber with perforated plates, which consists of fans for the supply of fresh air and an electric resistance heater. The surface air velocity was measured with an anemometer (HoldPeak® model HP 866B, wind speed accuracy ± 5%, Zhuhai, China), digital thermometer (Extech Instruments, model 421305, Waltham, Massachusetts, USA), air humidity is controlled with a digital thermohygrometer (BOE 330), and drying time with a digital stopwatch (Huawei Band 6-7E5). During all hot air-drying experiments, the airflow direction was parallel to the samples. The microbiological analysis was performed on all flours obtained by milling the dry pods and then sieving them to a particle size of 100 μm.

2.2. Analysis

Moisture, total solids in flour, mesophilic aerobic count, mould and yeast count, *E. coli* counts, and *S. aureus* were determined by AOAC official methods [6]. *Salmonella* spp. was determined by ISO 6579-1:2017 [7] by counting on a plate. Microbiological parameters such as mesophilic aerobic count, yeast and mould, count *Escherichia coli count*, *Staphylococcus aureus* count, and the absence of *Salmonella* spp. were confirmed; each of the determinations was made in triplicate. The result of the counts was expressed in CFU (colony-forming units)/g. In the case of *Salmonella* spp., the result was expressed as absence/25 g.

3. Results and Discussion

Method of Disinfection and Microbiological Counting of Carob Flours

In Table 1, the microbiological results of the samples of carob pod flour subjected to rinsing with different disinfectants obtained by drying in trays with forced circulation air at 60 °C for 7 h are presented. According to the Peruvian Technical Standard NTP 209.602 Carob Flour: Definitions and Requirements [5], the maximum value of mesophilic aerobic count, mould and yeast count, *E. coli* counts, and *S. aureus* counts is 1×10^2 CFU/g, and for *Salmonella*, absence is required. The counts of mesophilic aerobic count applying disinfection of 3% citric acid and 0.3% hypochlorite are within the parameter established by the standard used, but not for the parameter of counting mould and yeast count and *Staphylococcus aureus* count, which applying the disinfection of 0.3% hypochlorite is not within the parameter established by the standard. The microbiological parameters for carob flours for mesophilic aerobic count, mould and yeast counts, *E. coli* count, *S. aureus* count, and *Salmonella* spp. [5] were in the acceptable range with treatment in a 3% citric acid solution.

Table 2 shows the results obtained for the determination of moisture in the flours obtained in all treatments. Drying maintained for 7 h achieved a humidity of less than 5% in all the samples analysed, with no significant differences in the means (ANOVA, Tuckey post-test $p < 0.05$).

This research proposes an effective method of disinfection to obtain carob flour (*Neltuma* sp.) with microbiological quality under local conditions, which must be adjusted for different species of carob trees that present morphological differences that may influence the drying process. This tradition of taking advantage of the various fruits from the Chaco mountain has deep roots in ancestral history. The carob flour can be an excellent alternative for gluten-free muffins and baked goods, where, in general, the technological quality is not affected in the resulting products (except for a smaller volume in the dough), but they improve in the extension of shelf life by reducing the retrogradation of starch as a result of the modification of the gluten protein matrix exerted by the dietary fibre and globular proteins of carob flour [8]. Thus, the carob fruits of the Paraguayan Chaco have a great potential for use at a nutritional and industrial level within the framework of Food and Nutritional Security and are necessary conservation programs and management plans for the sustainability of this food resource.

Table 1. Results of the determination of mesophilic aerobes, *Escherichia coli*, *Staphylococcus aureus*, *Salmonella* spp., moulds, and yeasts in samples of carob pod meal.

Variables	Mesophilic Aerobes (CFU/g)	Moulds and Yeasts (CFU/g)	*E. coli* (CFU/g)	*Staphylococcus aureus* (CFU/g)	*Salmonella* spp. (CFU/g)
Citric Acid 3%	2×10	<10	<10	1×10	Absence/25 g
Citric Acid 1%	1×10^4	<10	<10	1×10^2	Absence/25 g
Sodium hypochlorite 0.3%	5×10	2.5×10^2	<10	1×10^3	Absence/25 g
Sodium hypochlorite 0.1%	1×10^3	3×10^2	<10	2×10^2	Absence/25 g
Rinse with Water	4×10^3	1×10^4	<10	3×10^3	Absence/25 g
No rinse/Undried	1×10^3	1×10^6	<10	1×10^4	Absence/25 g
Max. value Reference (CFU/g)	1×10^2	1×10^2	1×10^2	1×10^2	Absence/25 g

Reference: Peruvian Technical Standard NTP [5]. CFU: Colony-Forming Units.

Table 2. Moisture at different drying times by trays with forced air circulation.

Time	Moisture (%)
2 h	7.03 ± 0.02
4 h	6.32 ± 0.03
6 h	5.48 ± 002
7 h	4.80 ± 0.03

The results are expressed as an average and their corresponding standard deviations (X ± DS).

4. Conclusions

Washing the whole pods with 3% citric acid was optimal for the count of microorganisms analysed: Mesophilic aerobes, fungi, and yeasts; *Escherichia coli*, *Staphylococcus aureus*, and *Salmonella* spp. It was observed that the combined process of disinfection and drying in trays with forced air circulation for 7 h at 60 °C has a positive effect on the microbiological quality of the flours obtained and complies with the humidity standards necessary for their marketing and consumption. This reinforces the position of carob as a viable raw material for the production of safe food.

Author Contributions: Conceptualization, L.M. and K.M.; methodology, K.M.; software, L.M.; validation, N.S., K.M. and R.V.; formal analysis, N.S. and R.V.; investigation, N.S.; resources, L.M.; data curation, K.M.; writing—original draft preparation, L.M.; writing—review and editing, K.M.; visualization, L.M.; supervision, S.C.; project administration, S.C.; funding acquisition, L.M. All authors have read and agreed to the published version of the manuscript.

Funding: This research was funded by *Consejo Nacional de Ciencia y Tecnología* (CONACYT) with the support of *Fondo para la Excelencia de la Educación e Investigación* (FEEI) grant number PINV01-168.

Institutional Review Board Statement: Not applicable.

Informed Consent Statement: Not applicable.

Data Availability Statement: Data are contained within the article.

Acknowledgments: This work was supported by Ia ValSe-Food-Network and Lic. Adeline Friesen (Tucos Factory E.I.R.L.).

Conflicts of Interest: The authors declare no conflict of interest.

References

1. Menna, F.; Bianco, V. *Frutos Silvestres Del Gran Chaco: Saberes y Recetas de Las Mujeres Indígenas Rescatados Por Un Baluarte Slow Food*; Slow Food: Bra, Italy, 2016.
2. Villalba, R.; Denis Ibars, J.; Martínez, K.; Coronel, E.; Friesen, A.; Mereles, L. Variations in the Composition of "Algarrobas" (*Prosopis* sp.) Flours from Paraguayan Chaco. In Proceedings of the IV Conference Ia ValSe-Food CYTED and VII Symposium Chia-Link, La Plata and Jujuy, Argentina, 14–18 November 2022; MDPI: Basel, Switzerland, 2022; p. 25.
3. U. S. Food and Drug Administration. Handling Flour Safely: What You Need to Know. Available online: https://www.fda.gov/food/buy-store-serve-safe-food/handling-flour-safely-what-you-need-know (accessed on 31 October 2024).
4. Mom, M.P.; Romero, S.M.; Larumbe, A.G.; Iannone, L.; Comerio, R.; Smersu, C.S.S.; Simón, M.; Vaamonde, G. Microbiological Quality, Fungal Diversity and Aflatoxins Contamination in Carob Flour (Prosopis Flexuosa). *Int. J. Food Microbiol.* **2020**, *326*, 108655. [CrossRef] [PubMed]
5. *Norma Técnica Peruana NTP 209.602*, Harina de Algarroba: Definiciones y Requisito, 1st ed.; Instituto Nacional de Defensa de la Competencia y de la Intelectual (INDECOPI): San Borja, Peru, 2007.
6. Horwitz, W.; Chichilo, P.; Reynols, H. *Official Methods of Analysis of the Association of Official Analytical Chemists*, 17th ed.; AOAC: Gaithersburg, MA, USA, 2000.
7. *ISO 6579-1:2017*; (Annex D) Microbiology of the Food Chain, Horizontal Method for the Detection, Enumeration and Serotyping of *Salmonella* spp. Part 1 Detection of *Salmonella* spp. ISO: Geneva, Switzerland, 2017.
8. Tsatsaragkou, K.; Mandala, I.; Stoforos, N.G. Fermentation Kinetics of Gluten-Free Breads: The Effect of Carob Fraction and Water Content. *Foods* **2023**, *12*, 1809. [CrossRef] [PubMed]

biology and life sciences forum

MDPI

Proceeding Paper

Development of Instant Puree from Native Potatoes (*Solanum andigenum*) and Black Mashua (*Tropaeolum tuberosum*) Fortified with Black Quinoa (*Chenopodium quinoa*) †

Pablo Castañeda * and Nancy Chasquibol

Carrera de Ingeniería Industrial, Universidad de Lima, Av. Javier Prado este 4600, Fundo Monterrico Chico, Surco, Lima 15023, Peru; nchasquibol@ulima.edu.pe

* Correspondence: pablo_castanedafalcon@hotmail.com

† Presented at the VI International Congress la ValSe-Food, Lima, Peru, 23–25 September 2024.

Abstract: In Peru's Andean region, a diversity of seeds and tubers with high nutritional value and health benefits are grown. Nevertheless, chronic malnutrition and obesity emphasize the need to take advantage of our agricultural wealth to improve public health and ensure sustainable development. The aim of this study was to develop an instant puree with native potatoes (*Solanum andigenum*) and black mashua (*Tropaeolum tuberosum*) fortified with black quinoa (*Chenopodium quinoa*). This study employed a simplex centroid mixture design. The proximal compositions of the three formulations developed were as follows: moisture content of 9.37 ± 0.13% to 9.45 ± 0.06%, ash content of 3.34 ± 0.02% to 3.79 ± 0.17%, protein content of 9.48 ± 0.25% to 11.16 ± 0.38% and total carbohydrate content of 72.81 ± 0.35% to 74.98 ± 0.22%. The samples showed significantly higher antioxidant (7280 ± 113.5 μg trolox/g powder to 12914 ± 604 μg trolox/g powder) and phenolic (3444 ± 241 μg gallic acid equivalent (GAE) /g powder to 7044 ± 322 μg GAE/g powder) content than the control sample. Also, the results of the techno-functional properties of the samples were as follows: water absorption capacity 3.56 ± 0.92 g H_2O/g to 3.95 ± 0.07 g H_2O/g, solubility 14.45 ± 0.07% to 17.88 ± 0.15% and in vitro protein digestibility 70.27 ± 0.05% to 71.61 ± 0.8%. The samples demonstrated an adequate balance of amino acids compared with the control sample. The sensory characteristics of rehydrated powders were determined. Therefore, in a fast-paced world where convenience food options are part of a continuously expanding market, a nutritionally improved instant puree from ancestral crops is not only more nutritious and tastier but also contributes to sustainability and promotes culinary diversity.

Keywords: Andean potato; black mashua; black quinoa; instant puree

1. Introduction

Peru is the leading potato producer in Latin America, with the more than 4.5 million tons per year produced being the livelihood of more than 700,000 Peruvian families, growing over 4000 varieties of native potatoes (*Solanum andigenum*). These varieties are grown in different regions of the country, especially in the Andes, where climatic and geographical diversity has encouraged the development of a wide range of potatoes with different colors, shapes, textures and flavors. Purple and red potato varieties are abundant in anthocyanins and flavonoids. Here, this tuber has an excellent culinary quality, and it has long been part of the basic diet of the Andean Peruvian population [1].

Mashua (*Tropaeolum tuberosum*) is a tuber originally from the Peruvian Andes that grows between 3500 and 4100 m above sea level, and it extends to countries such as Bolivia, Colombia and Ecuador. Among Andean tubers, it boasts one of the highest yields, with productivity reaching up to 70 tons per hectare, and it is one of the simplest to cultivate. For centuries, this tuber has been utilized in traditional medicine for various ethnic groups in the Andean regions of South America and it is considered a valuable source for developing functional foods due to its impressive nutritional and health-promoting properties. It possesses a significant diversity in shapes and colors, varying from yellowish-white to dark reddish-purple. This last variety has eight to ten times higher antioxidant activity than yellow-colored varieties [2].

Quinoa (*Chenopodium quinoa*) is an Andean seed that is considered an excellent source of macro- and micronutrients. This gluten-free crop possesses an excellent amino acid composition and bioactive compounds, which have been recognized for their significant health benefits, including their roles as antioxidants, hypolipidemic agents, antidiabetic agents, anti-inflammatory agents, and anticancer agents. Darker quinoa seeds contain higher levels of phenolic compounds and have greater antioxidant properties [3].

Over recent years, plant-based convenience foods have become increasingly important in modern diets because of the need to consume easy-to-prepare and time-saving healthy meals. Instant puree products are part of a continuously expanding market of convenience food options. This research centers on the development and characterization of an instant puree from native potatoes (*Solanum andigenum*) and black mashua (*Tropaeolum tuberosum*) fortified with black quinoa (*Chenopodium quinoa*).

2. Materials and Methods

2.1. Raw Material

The Andean potato Azul (*Solanum andigenum*) and mashua (*Tropaeolum tuberosum*) were acquired from Condorcocha Huamanga, Ayacucho, and quinoa (*Chenopodium quinoa*) was purchased from a local market in Lima, Peru. The control sample was a commercial instant puree that was obtained from a local market in Lima, Peru.

2.2. Design of Experiment (DOE)

The design of the experiment was a simplex centroid mixture design using Minitab 19 software (Stat-Ease, Inc., Minneapolis, MN, USA). Two independent variables were used: X1—mashua (10–40%) content and X2—potato (40–70%) content. The protein (%) content, polyphenol (µg GAE/g powder) content and antioxidant activity (µg Trolox/g powder) were determined as the response variables, as shown in Table 1.

Table 1. Design of experiment (DOE).

Formulations	Std Order	Run Order	PtType	Blocks	Andean Potato	Mashua	Protein (%)	TPC (µg GAE/g Powder)	DPPH (µg Trolox/g Powder)
F1	5	1	−1	1	0.7	0.1	9.48 ± 0.25	3445 ± 241	12,915 ± 604
F2	1	2	1	1	0.4	0.4	11.16 ± 0.38	7044 ± 322	7281 ± 113
F3	2	3	1	1	0.625	0.175	9.91 ± 0.37	4651 ± 223	11,877 ± 659
F4	4	4	−1	1	0.55	0.25	10.25 ± 0.28	5543 ± 253	10,755 ± 259
F5	3	5	0	1	0.475	0.325	10.70 ± 0.35	6231 ± 271	8842 ± 193

F1: formulation 1; F2: formulation 2; F3: formulation 3; F4: formulation 4; F5: formulation 5; TPC: total phenolic content.

2.3. Sample Preparation

Formulations of dehydrated purees were developed with different amounts of native potato (*Solanum andigenum*) and mashua (*Tropaeolum tuberosum*), keeping the quinoa (*Chenopodium quinoa*) content constant. The tubers (potato and mashua) were washed, cut into thin slices and cooked at 80 °C for 9 min. The quinoa, previously washed, was cooked at 80 °C for 8 min. All three samples were dehydrated by an infrared dryer (IRC D18, Irconfort, Seville, Spain) at 40 °C for 16 h, 24 h and 12 h, respectively. Then, the samples were ground in a food shredder (Grindomix GM200, Restch, Haan, Germany) and sieved (300 µm) to obtain a fine powder to finally be mixed in proportions according to Table 1 and stored in polyethylene bags at room temperature for subsequent analysis.

2.4. Characterization

2.4.1. Proximal Composition

This analysis was performed using official methods. The moisture content was determined at 110 °C to a constant weight, the ash content was determined by the ignition method (550 °C for 72 h), and the fat content was determined with hexane for 9 h. The total protein (% nitrogen × 6.25) was analyzed by a Kjeldahl analyzer (UDK 139, VELP, Usmate Velate, Italy).

2.4.2. Determination of Mineral Content

The mineral content (Ca, Cu, Fe, K, Mg, Na, P, Zn) was determined by atomic absorption spectrophotometry (Nexion 350x, Perkin Elmer, Waltham, MA, USA) [4].

2.4.3. Determination of Amino Acid Profile

The amino acid profile was quantified by HPLC (ARC, Waters, Milford, CT, USA) [5], with a 150 mm × 3.9 mm inner diameter C18 reverse-phase column. All measures were performed in duplicate.

2.4.4. Antioxidant Activity

The antioxidant activity was determined by spectrometry with the 2,2-diphenyl-1-picrylhydrazyl (DPPH) method at 517 nm, and with the 2,2′-Azino-bis (3-ethylbenzthiazoline-6-sulfonic acid) (ABTS) method at 734 nm (UV 1280 Vis Spectrophotometer Shimadzu, Kyoto, Japan) [6]. The results were expressed as μg Trolox/g powder and were performed in triplicate.

2.4.5. Total Phenolic Content (TPC)

The TPC was determined by the Folin–Ciocalteau method [6] at 760 nm (UV 1280 Vis Spectrophotometer Shimadzu, Kyoto, Japan). The results were performed in triplicate and were expressed as the μg of gallic acid equivalent (GAE)/g powder.

2.4.6. In Vitro Protein Digestibility

This was determined by the method of Tapia et al. [6] with slight modification. Analyses were performed in duplicate, and the results are expressed as:

$$\text{IVPD (Digestibility (\%)} = 65.66 + 18.10 \times (\text{pH 0 min} - \text{pH 10 min}))$$

2.4.7. Solubility

The solubility was determined by dissolving 0.5 g of powder in 20 mL of distilled water in a volumetric flask, followed by vortexing for 5 min and centrifuging at 3000 rpm for 5 min. In total, 20 mL of the supernatant was taken and heated to 105 °C for 2 h. Solubility (%) was calculated by weight difference [4].

2.4.8. Water Holding Capacity (WAI)

This was determined by dissolving 1 g of sample in 50 mL of distilled water in a centrifuge tube and leaving to stand for 30 min at room temperature followed by centrifuging (HettichZentrifugen-Mikro, München, Germany) for 15 min at 4000 rpm. The WAI was determined as the weight of the gel obtained after removal of the supernatant per unit weight of original dry solids [4].

2.4.9. Sensory Analysis

A panel of 30 university students between 18 and 26 years old evaluated the appearance, flavor, texture and overall acceptability. The test was based on a 9-point hedonic scale (1 = dislike extremely and 9 = like extremely).

2.5. Statistical Analysis

Analysis of variance (ANOVA) was analyzed at a 95% significance level with Minitab 19.0 Software (Minitab Inc., State College, Palo Alto, CA, USA).

3. Results and Discussion

According to Table 1, F2 presented the highest protein content (11.16 ± 0.38%) and the highest amount of TPC (7044 ± 322 µg GAE/g powder), while F1 presented the highest antioxidant activity by DPPH (12,915 ± 604 µg Trolox/g dm), followed by F3 with 11,877 ± 659 µg Trolox/g dm, which also presented a high protein content (9.91 ± 0.37%). For these reasons, F1, F2 and F3 were chosen for the subsequent analysis.

The proximate composition is presented in Table 2. All three formulations showed higher protein content than the control sample, as well as in comparison with the other commercial brands in Peru and abroad (7.14% to 8.4% protein content). Formulation 2 presented higher protein (11.06 ± 0.38%) content than the dehydrated puree (9.92%) from native potato (*Solanum andigenum*) and oca (*Oxalis tuberosa*) fortified with quinoa (*Chenopodium quinoa*), which had 40%, 30% and 30% content, respectively, as reported by Inca [7]. Also, this last puree presented less fat content than the formulations (2.46%). The control sample showed more carbohydrate (77.89 ± 0.33%) content and less fat (0.95 ± 0.06%) content than the formulations. The moisture content had 9.5% accordance with the Peruvian regulation NTP 011.808.

Table 2. Proximal composition of instant purees from native potatoes (*Solanum andigenum*) and black mashua (*Tropaeolum tuberosum*) fortified with black quinoa (*Chenopodium quinoa*).

Sample	Moisture (%)	Ash (%)	Protein (%)	Fat (%)	Carbohydrates (%)
CS	9.40 ± 0.12 A	3.76 ± 0.36 A	8.1 ± 0.4 C	0.95 ± 0.06 D	77.9 ± 0.3 A
F1	9.45 ± 0.06 A	3.34 ± 0.02 B	9.48 ± 0.25 BC	2.75 ± 0.01 C	75.0 ± 0.2 B
F2	9.33 ± 0.07 A	3.76 ± 0.04 B	11.16 ± 0.38 A	2.94 ± 0.04 A	72.8 ± 0.3 D
F3	9.37 ± 0.13 A	3.79 ± 0.17 A	9.91 ± 0.37 B	2.84 ± 0.02 B	74.1 ± 0.3 C

Results are expressed as means ± SD (n = 3). A, B, C and D values in the same column with different letters differ significantly when $p < 0.05$.

The results of TPC, DPPH, in vitro protein digestibility, solubility and water holding capacity are shown in Table 3. All formulations presented a higher amount of TPC (3445 ± 241 µg GAE/g powder to 7044 ± 322 µg GAE/g powder) than the control sample (1071 ± 113 µg GAE/g powder), increasing according to the amount of mashua (17,387 ± 356 µg GAE/g powder) in the sample. The antioxidant activity according to DPPH (7281 ± 113 µg Trolox/g powder to 12,915 ± 604 µg Trolox/g powder) was higher than the control sample (5654 ± 188 µg Trolox/g powder), increasing according to the amount of native potato (15,469 ± 307.7 µg Trolox/g powder) in the sample. All three formulations showed higher in vitro protein digestibility (70.27 ± 0.38% to 71.61 ± 0.8%) and solubility (14.45 ± 0.34% to 17.88 ± 0.15%) than the control sample (7.64 ± 0.13%) However, all three formulations presented a lower water holding capacity (3.56 ± 0.09 g/g to 3.95 ± 0.07 g/g) than the control sample (7.5 ± 0.16 g/g).

The composition of the amino acid profiles is summarized in Table 4. Compared to the control sample, the predominant amino acids were threonine (34.3 ± 1.6 to 46.6 ± 1.0), valine (42.5 ± 0.3 to 55.9 ± 1.0) and phenylalanine (30.3 ± 1.4 to 44.4 ± 0.9). The three formulations showed a balanced profile of essential amino acids according to the FAO/WHO recommendations. These formulations showed higher contents of specific amino acids: leucine (70.3 ± 1.7 to 91.9 ± 0.6), lysine (36.8 ± 1.6 to 56.1 ± 1.5) and tryptophan (14.9 ± 0.2 to 21.39 ± 0.84). Quinoa (*Chenopodium quinoa*) provided amino acids, especially leucine and lysine, to the purees. This last amino acid supports human growth and boosts immune function, making it a crucial nutrient for children [3]. The puree formulations contained high concentrations of tryptophan, more than double and triple the recommended amount according to the FAO/WHO.

Table 3. TPC, DPPH, in vitro digestibility, solubility and water holding capacity of instant purees made from native potatoes (*Solanum andigenum*) and black mashua (*Tropaeolum tuberosum*) fortified with black quinoa (*Chenopodium quinoa*).

Sample	TPC (µg GAE/g Powder)	DPPH (µg Trolox/g Powder)	In Vitro Protein Digestibility (%)	Solubility (%)	Water Holding Capacity (g/g)
CS	1071 ± 113.5 ^{D}	5654 ± 188,3 ^{C}	68.21 ± 0.74 ^{B}	7.64 ± 0.13 ^{D}	7.5 ± 0.16 ^{A}
F1	3445 ± 241.5 ^{C}	12,915 ± 604 ^{A}	71.61 ± 0.8 ^{A}	14.45 ± 0.34 ^{C}	3.56 ± 0.09 ^{C}
F2	7044 ± 322.6 ^{A}	7281 ± 113.5 ^{B}	70.27 ± 0.38 ^{A}	17.88 ± 0.15 ^{A}	3.95 ± 0.07 ^{B}
F3	4651 ± 223.4 ^{B}	11,877 ± 659 ^{D}	70.93 ± 0.75 ^{A}	16.21 ± 0.22 ^{B}	3.73 ± 0.06 ^{C}

TPC: total phenolic content, DPPH: 2,2-diphenyl-1-picrylhydrazyl. Results are expressed as means ± SD (n = 3). A, B, C and D values in the same column with different letters differ significantly when $p < 0.05$.

Table 4. Profiles of essential amino acids in instant purees made from native potatoes (*Solanumandigenum*) and black mashua (*Tropaeolum tuberosum*) fortified with black quinoa (*Chenopodium quinoa*).

Amino Acids	CS	F1	F2	F3	FAO%
	(mg Amino Acid/g Protein)				
Aspartic acid	124.1 ± 1.17 ^{B}	139.9 ± 0.5 ^{A}	104.4 ± 1.7 ^{D}	117.7 ± 1.9 ^{C}	
Glutamic acid	84.3 ± 1.67 ^{D}	171.5 ± 1.8 ^{A}	124.9 ± 1.4 ^{C}	145 ± 28 ± 1.6 ^{B}	
Serine	21.1 ± 1.83 ^{C}	45.78 ± 0.1 ^{A}	38.2 ± 1.4 ^{B}	43.1 ± 1.6 ^{A}	
Histidine	1.58 ± 0.44 ^{C}	20.4 ± 1.3 ^{A}	16.1 ± 1.4 ^{B}	18.2 ± 1.6 ^{AB}	15
Glycine	22.8 ± 1.8 ^{D}	47.5 ± 1.0 ^{A}	32.3 ± 1.7 ^{C}	36.4 ± 1.9 ^{B}	
Threonine	21.5 ± 1.6 ^{D}	46.6 ± 1.0 ^{A}	34.3 ± 1.6 ^{C}	38.6 ± 1.8 ^{B}	23
Arginine	46.2 ± 0.9 ^{A}	21.1 ± 1.6 ^{B}	19.8 ± 0.4 ^{B}	21.8 ± 0.2 ^{B}	
Alanine	19.1 ± 1.9 ^{D}	59.1 ± 0.9 ^{A}	36.8 ± 1.6 ^{C}	41.5 ± 1.8 ^{B}	
Proline	25.3 ± 1.5 ^{D}	64.2 ± 1.8 ^{A}	48.4 ± 1.2 ^{C}	54.6 ± 1.4 ^{B}	
Tyrosine	18.2 ± 1.1 ^{D}	37.1 ± 1.2 ^{A}	25.8 ± 0.5 ^{C}	29.1 ± 0.6 ^{B}	38
Ammonia	32.6 ± 1.2 ^{C}	43.2 ± 1.7 ^{A}	29.0 ± 1.4 ^{C}	32.7 ± 1.6 ^{C}	
Valine	26.7 ± 1.5 ^{D}	55.92 ± 1.0 ^{A}	42.5 ± 0.3 ^{C}	48.0 ± 0.32 ^{B}	39
Methionine	6.34 ± 0.2 ^{C}	11.9 ± 1.1 ^{B}	14.1 ± 1.1 ^{A}	13.1 ± 0.3 ^{AB}	22
Cysteine	1.9 ± 0.2 ^{C}	3.8 ± 0.09 ^{A}	2.96 ± 0.3 ^{B}	3.4 ± 0.4 ^{AB}	
Isoleucine	17.9 ± 1.4 ^{D}	38.7 ± 1.01 ^{A}	27.94 ± 1.0 ^{C}	31.5 ± 1.1 ^{B}	30
Leucine	49.7 ± 1.6 ^{D}	91.9 ± 0.6 ^{A}	70.30 ± 1.7 ^{C}	79.3 ± 1.9 ^{B}	59
Phenylalanine	25.5 ± 1.8 ^{D}	44.35 ± 0.9 ^{A}	30.32 ± 1.40 ^{C}	35.9 ± 1.6 ^{B}	
Lysine	27.94 ± 1.64 ^{D}	56.08 ± 1.5 ^{A}	36.75 ± 1.63 ^{C}	44.8 ± 1.8 ^{B}	45
Tryptophan	5.93 ± 0.57 ^{D}	14.88 ± 0.23 ^{C}	21.39 ± 0.84 ^{A}	19.3 ± 0.1 ^{B}	6

Results are expressed as means ± SD (n = 3). A, B, C and D values in the same column with different letters differ significantly when $p < 0.05$.

Total minerals are presented in Table 5. Formulation 2 presented a higher quantity of minerals than the control: calcium (1271 ± 101.2 mg/kg), iron (35.41 ± 2.83 mg/kg), potassium (11,092 ± 887 mg/kg), magnesium (1462 ± 116.9 mg/kg), phosphorus (3353.82 ± 260.3 mg/kg) and zinc (21.02 ± 1.68 mg/kg). According to the daily requirements for children between 1 and 3 years old given by the National Institute of Health [8], instant puree (F2) would cover about 15.17% of the daily requirement of iron. Therefore, the puree developed could contribute to reductions in malnutrition and childhood anemia in our country.

In the sensory evaluation, on a scale of 1 to 9, the characteristics of appearance, color, smell, texture and taste had average values of 7.3, 7.0, 8.0, 6.0 and 6.0, respectively. These results show that the puree was liked by the panelists.

Table 5. Total minerals of instant purees made from native potatoes (*Solanum andigenum*) and black mashua (*Tropaeolum tuberosum*) fortified with black quinoa (*Chenopodium quinoa*).

Minerals (mg/kg)	CS (mg/kg)	F2 (mg/kg)
Calcium	282 ± 22 [b]	1271 ± 101 [a]
Copper	1.62 ± 0.08 [b]	3.12 ± 0.20 [a]
Iron	9.89 ± 0.62 [b]	35.4 ± 2.8 [a]
Potassium	11,489 ± 896 [a]	11,092 ± 887 [b]
Magnesium	678 ± 51 [b]	1462 ± 116 [a]
Sodium	361.0 ± 28.8 [a]	168.2 ± 13.4 [b]
Phosphorus	1517 ± 121 [b]	3354 ± 260 [a]
Zinc	6.95 ± 0.55 [b]	21.0 ± 1.6 [a]

Results are expressed as means ± SD (n = 3). a, b, c and d values in the same column with different letters differ significantly when $p < 0.05$.

4. Conclusions

The use of native Andean tubers instead of commercial tubers such as white and yellow potatoes showed significantly higher antioxidant activity and total phenolic content. Formulation 2 presented the highest amount of TPC 7044.2 ± 322 (µg GAE/g powder), being seven times higher than that of the control sample and presented the highest protein (11.2 ± 0.3%) content. Formulation 1 presented the highest antioxidant activity according to DPPH (12,915 ± 604 µg Trolox/g powder). The addition of quinoa not only increased the protein content but also provided a balanced profile of essential amino acids. The puree was accepted by the individuals, and in comparison to commercial instant purees, the formulations were more nutritious. Therefore, in a continuously expanding market of convenience food options, the instant puree developed is an alternative for the modern diet.

Author Contributions: Conceptualization, P.C. and N.C.; methodology, P.C. and N.C.; software, P.C.; validation, P.C. and N.C.; formal analysis, P.C. and N.C.; investigation, P.C. and N.C.; resources, P.C.; data curation, P.C. and N.C.; writing—original draft preparation, P.C. and N.C.; writing—review and editing, P.C. and N.C.; visualization, P.C. and N.C.; supervision, P.C. and N.C.; project administration, P.C. and N.C. All authors have read and agreed to the published version of the manuscript.

Funding: This research received no external funding.

Institutional Review Board Statement: Not applicable.

Informed Consent Statement: Not applicable.

Data Availability Statement: Data are contained within the manuscript.

Acknowledgments: This work was developed at Centro de Alimentos Funcionales of Carrera de Ingenieria Industrial and supported by Universidad de Lima, Peru.

Conflicts of Interest: The authors declare no conflicts of interest.

References

1. Bellumori, M.; Silva, N.A.C.; Vilca, L.; Andrenelli, L.; Cecchi, L.; Innocenti, M.; Balli, D.; Mulinacci, N. A Study on the Biodiversity of Pigmented Andean Potatoes: Nutritional Profile and Phenolic Composition. *Molecules* **2020**, *25*, 3169. [CrossRef] [PubMed]
2. Ticona, L.N.A.; Pérez, V.T.; Benito, P.B. Local/traditional uses, secondary metabolites and biological activities of Mashua (*Tropaeolum tuberosum* Ruiz & Pavon). *J. Ethnopharmacol.* **2020**, *247*, 112152. [CrossRef]
3. Ren, G.; Teng, C.; Fan, X.; Guo, S.; Zhao, G.; Zhang, L.; Liang, Z.; Qin, P. Nutrient composition, functional activity and industrial applications of quinoa (*Chenopodium quinoa* Willd.). *Food Chem.* **2023**, *410*, 135290. [CrossRef] [PubMed]
4. Chasquibol, N.; Sotelo, A.; Alarcón, R. Development of Powdered Beverage with Cushuro (*Nostoc commune*) Concentrated Protein and Quinoa (*Chenopodium quinoa*). *Biol. Life Sci. Forum* **2023**, *25*, 2. [CrossRef]
5. Chasquibol, N.; Gonzales, B.F.; Alarcón, R.; Sotelo, A.; Márquez-López, J.C.; Rodríguez-Martin, N.M.; del Carmen Millán-Linares, M.; Millán, F.; Pedroche, J. Optimisation and Characterisation of the Protein Hydrolysate of Scallops (*Argopecten purpuratus*) Visceral By-Products. *Foods* **2023**, *12*, 2003. [CrossRef] [PubMed]
6. Tapia, M.; Marimón, S.J.; Salazar, N. Development of Extruded Snacks with Protein Hydrolysed from Jumbo Squid (*Dosidicus gigas*) by-Product and Cañihua (*Chenopodium pallidicaule Aellen*). *Biol. Life Sci. Forum* **2023**, *25*, 4. [CrossRef]

7. Inca, E. Evaluacion las Propiedades Tecnofuncionales y Sensoriales de Pure Deshidratado Formulado con papa Nativa Yawar Huayco (*Solanum tuberosum*) Fortificado con Quinua Variedad Blanca de Junín (*Chenopodium quinoa Willd*) y Oca variedad Zapallo (*Oxalis tuberosa Mol*). [Thesis to Obtain a Professional Degree of Agroindustrial Engineer, Pontificia Universidad Javeriana]. 2015. Available online: https://hdl.handle.net/20.500.14168/212 (accessed on 23 October 2024).
8. National Institutes of Health. Hierro. Hoja Informativa para Consumidores. Available online: https://ods.od.nih.gov/factsheets/Iron-DatosEnEspanol/ (accessed on 1 July 2024).

biology and life sciences forum

MDPI

Proceeding Paper

Development of High-Protein Cookies Enriched with Defatted Sacha Inchi (*Plukenetia huayllabamana*) Cake and Tarwi (*Lupinus mutabilis Sweet*) to Combat Child Malnutrition †

Mateo Tapia [1,*], Nicolás Diez [1], Axel Sotelo [1], Rafael Alarcón [1] and Nelson Tapia [2]

[1] Carrera de Ingeniería Industrial, Universidad de Lima, Av. Javier Prado Este 4600, Surco, Lima 15023, Peru; 20213508@aloe.ulima.edu.pe (N.D.); aasotelo@ulima.edu.pe (A.S.); ralarcor@ulima.edu.pe (R.A.)

[2] Facultad de Química e Ingeniería Química, Universidad Nacional Mayor de San Marcos, Av. Carlos Germán Amezaga 375, Lima 15023, Peru; ntapia@unmsm.edu.pe

* Correspondence: mtapiac@ulima.edu.pe

† Presented at the VI International Congress la ValSe-Food, Lima, Peru, 23–25 September 2024.

Abstract: In Peru, 5 of every 10 children suffer from malnutrition, which increases the risk of developing non-communicable diseases (NCDs) at the early stages of their life. Sacha inchi (*Plukenetia huayllabamana*) cake is the by-product of the oil extraction process of the Amazon Peruvian seed and has high protein (56%) and fiber (5.9%) content. Tarwi (*Lupinus mutabilis Sweet*) is a Peruvian legume characterized by its high protein content (>50%) and high levels of tyrosine and tryptophan amino acids. Thus, the purpose of this research was to develop high-protein cookies enriched with defatted sacha inchi cake (DSIC) and defatted tarwi (DT) as an alternative snack to combat child malnutrition. Five types of cookies were developed, along with one control sample, containing corn flour (50% *w/w*), rice flour (20% *w/w*), and varying the content of quinoa flour, DSIC flour and DT flour (0, 10%, 30% *w/w*). The addition of DSIC and DT increased the content in protein from 12.13% to 20.30%, fat from 6.76% to 10.42%, and ash from 1.51% to 1.89% compared to the control sample. In contrast, the moisture content decreased from 10.21% to 10.78% and carbohydrates from 59.73% to 67.48%. Moreover, all cookies had high in vitro digestibility (70.91% to 74.17%) and relatively high antioxidant activity (346 to 469 μg GAE/g cookie) and phenolic content (1286 to 1755 μg trolox/g cookie). Sensory analysis showed that cookies enriched with DT were more appealing to the panelists than those enriched with DSIC and the control sample. Hence, these cookies could serve as a gluten-free nutritional food alternative to combat child malnutrition.

Keywords: cookies; tarwi; sacha inchi; protein; nutritional value

1. Introduction

Peru is renowned worldwide for its natural biodiversity and rich culinary tradition. Nevertheless, 5 out of every 10 children suffer from malnutrition, defined as insufficient or excessive nutrient intake or an imbalance of essential nutrients, which can increase the risk of developing non-communicable diseases (NCDs), such as anemia, heart disease, diabetes, and others, at early stages of their lives [1]. Sacha inchi (*Plukenetia huayllabamana*), known as the "Giant Inca Peanut", is an Amazonian seed grown in the Amazonas region of Peru. It has an oil yield of 30.3% to 41.2% and a high omega-3 content (55.62% to 60.42%) [2]. However, a significant amount of by-product is generated after the oil extraction process, known as "cake", which accounts for around 40% to 50% of the total weight. This by-product contains approximately 56% protein, 5.19% fiber, 24.36% carbohydrates, and high amounts of tryptophan (44 mg/g protein) and phenylalanine (3 mg/g protein) [3]. Tarwi or "chocho" (*Lupinus mutabilis Sweet*) is a Peruvian legume domesticated and cultivated at altitudes between 1500 and 3850 m above sea level. It is characterized by its high protein content (>50%), fat content (~24%), low carbohydrate content (<10%), and high levels of tyrosine and tryptophan compared to other legumes [4]. Cookies are highly consumed by children and young adults due to their unique flavor, small compact size, texture, low cost, high-energy value, and practicality [5,6]. However, the primary ingredient of traditional cookies is wheat, which contains high amounts of gluten and can lead to adverse health effects in people that suffer wheat allergies, celiac disease, and non-celiac gluten sensitivities [6]. Hence, the purpose of this research was to develop high-protein cookies enriched with defatted sacha inchi cake (DSIC) and defatted tarwi (DT) as an alternative gluten-free snack to combat child malnutrition.

2. Materials and Methods

2.1. Raw Material

Sacha inchi (*Plukenetia huayllabambana*) seeds were collected in the province of Rodriguez de Mendoza, Department of Amazonas, Peru, and cold-pressed at the Laboratorio de Alimentos Funcionales at Universidad de Lima. Afterward, the cake was separated, grounded in a food shredder (Grindomix GM200, Retsch, Haan, Germany), and stored in aluminized bags. Meanwhile, tarwi (*Lupinus mutabilis Sweet*) flour was bought from a local market in Lima. Both flours were defatted using petroleum ether in a ratio of 1:6 (w/v) for 24 h. Afterward, they were left to dry at room temperature and stored in aluminum bags until further use. Corn flour, rice flour, quinoa flour, baking powder, butter, sugar, salt, eggs, and vanilla essence were bought from a local market in Lima, Peru.

2.2. Cookie Preparation

Five types of cookies (PC1–PC5) were developed with different amounts of DSIC and DT, plus one control, as shown in Table 1.

Table 1. Formulation of high-protein cookies (PC) enriched with defatted sacha inchi cake (DSIC) and defatted tarwi (DT).

Samples	Corn Flour (%)	Rice Flour (%)	Quinoa Flour (%)	DSIC Flour (%)	DT Flour (%)
Control	50	20	30	0	0
PC1	50	20	20	10	0
PC2	50	20	0	30	0
PC3	50	20	20	0	10
PC4	50	20	0	0	30
PC5	50	20	10	10	10

All cookies had the following ingredients: 200 g of flour mix (proportional to Table 1), 2.8 g of baking powder, 16 g of butter, 20 g of sugar, 2 g of salt, 3 egg whites, and 1 tablespoon of vanilla essence. For preparation, the method of Alexa et al. [7] was used with slight modifications. Consequently, the flour mix was homogenized using a planetary mixer (FPSTSMPL1-053, Oster, Guangdong, China), and the rest of the ingredients were added afterward and mixed again. Once a homogenous dough was obtained, it was placed in an aluminum bowl and cooled at 4 °C for 30 min. Subsequently, the dough was rolled out to a thickness of 8 cm, cut with a cookie cutter, and baked in a convection oven (HEB60R, Imaco, Guandong, China) at 180 °C for 15 min. After baking, the cookies were cooled and kept in aluminized bags.

2.3. Physical Properties

The water activity (WA) was determined at 25 °C using food water activity meter (VTS-160A, VTSYIQI, Shushan, China) and was performed in triplicate. For all cookies, the average weight (g) was determined, while the diameter (mm) and thickness (mm) were measured using a vernier caliper. All measurements were taken 12 times per batch of cookies [5].

2.4. Proximate Composition

The total protein content of the cookie was calculated as % nitrogen × 6.25 using a Kjeldahl analyzer (UDK 139, VELP, Usmate Velate, Italy) following official methods. The moisture content was assessed at 110 °C at a constant weight; the ash content was measured using the ignition method (550 °C for 72 h); the lipid content was determined with petroleum ether extraction for 6 h [8]. The energy values were obtained based on Koh et al. [6] and expressed in kcal.

2.5. Determination of Amino Acid Profile

The amino acid profile was estimated according to Chasquibol et al. [9] with slight modifications. The analysis was conducted using HPLC (ARC, Waters, Milford, CT, USA) with a 150 mm × 9 mm inner diameter C18 reverse-phase column. Tryptophan quantification was performed via HPLC following basic hydrolysis. All analyses were conducted in triplicate.

2.6. Determination of Total Phenolic Content (TPC)

TPC was determined by the Folin–Ciocalteu method [2] at 760 nm using a spectrophotometer (UV 1280 Vis Spectrophotometer Shimadzu, Kyoto, Japan). The results were expressed as µg of gallic acid equivalent (GAE)/g cookie. All analyses were performed in triplicate.

2.7. Antioxidant Activity

Antioxidant activity was determined by the DPPH method [2] with some modifications at 517 nm by spectrometry (UV 1280 Vis Spectrophotometer Shimadzu, Kyoto, Japan) and expressed as µg Trolox/g cookie. All analyses were performed in triplicate.

2.8. In Vitro Protein Digestibility (IPVD)

In vitro protein digestibility was measured by the method of Tinus et al. [8] with light modifications. The results were expressed as follows:

$$\text{IVPD } (\%) = 65.66 + 18.10 \times (\text{pH } 0 \text{ min} - \text{pH } 10 \text{ min})$$

2.9. Oxidative Stability

The oxidative stability was determined using an 892 Professional Rancimat© (Metrohm, Herisau, Switzerland) [2]. To calculate the shelf life at 25 °C, induction periods at 80, 120, and 150 °C were determined. Additionally, the acid index (NTP 206.013) was determined to evaluate the oxidation state of the samples at room temperature. The determinations were performed in duplicate.

2.10. Sensory Analysis

Forty panelists, between the ages of 18 and 35, assessed the sensory attributes of appearance, color, flavor, texture, aroma, and overall acceptability of the cookies. A 9-point hedonic scale was employed, with 1 being "dislike extremely" and 9 being "like extremely." The panelists were given 6 random samples in plastic cups, each marked with a distinct 3-digit code [8].

2.11. Statistical Analysis

The results were expressed as mean ± standard deviation, all measurements were conducted in triplicate, except for the measurement of oxidative stability and the IVPD, which were performed in duplicate. Analysis of variance (ANOVA) was used to analyze the acquired data at a 95% significance level (Minitab Inc., State College, PA, USA).

3. Results and Discussion

3.1. Physical Characteristics

All high-protein cookies showed a lower water activity (WA), ranging from 0.65 ± 0.24 to 0.68 ± 0.28, than the control sample (0.69 ± 0.16). Moreover, this WA is much higher than cookies reported by Sławińska et al. [5], 0.404 to 0.482, due to a higher amount of moisture in the high-protein cookies with DSIC and DT. Furthermore, all cookies had an approximate weight between 5.86 ± 0.14 g and 6.12 ± 0.13 g, with a diameter of 40.32 ± 0.09 to 40.45 ± 0.07 mm and thickness of 8.12 ± 0.12 mm to 8.44 ± 0.13 mm.

3.2. Proximate Composition

Analysis of the results in Table 2 indicates that the cookies containing DSIC and DT had higher contents of protein, lipids, and ash compared to the control sample. However, the control sample had higher moisture and carbohydrate content. Among the cookies, PC4 exhibited the highest protein content at 20.30 ± 0.13%, surpassing cookies made with *Agaricus bisporus* powder (6.67% to 7.20%), *Pleurotus ostreatus* powder (6.45% to 6.78%), lupin sprout flour (6.7% to 5.8%), and lupin green flour (7.5% to 11.3%) [5,7]. This trend was also observed for lipids and ash content. PC2 had the highest lipid content at 10.42 ± 0.28%, followed by PC4 at 8.33 ± 0.26%, whereas the control sample had the lowest lipid content at 5.25 ± 0.35%. Regarding ash content, PC2 had the highest value at 1.89 ± 0.04%, compared to 1.51 ± 0.06% in the control sample. In contrast, moisture and carbohydrate contents were lower in the samples containing DSIC and DT, primarily due to the higher protein and lipid contents in these flours. PC4 had the lowest moisture content at 10.21 ± 0.13% and the lowest carbohydrate content at 59.73 ± 0.54% among all cookies. The enriched cookies had energy values ranging from 390 ± 24 kcal (PC5) to 403 ± 21 kcal (PC2). Although these values are higher than the control sample (377 ± 40 kcal), the enriched cookies also have higher protein and ash contents and lower carbohydrate levels, making them more beneficial for daily nutrition. Additionally, PC5 provided a balanced profile, with a good amount of protein, lipids, carbohydrates, ash, and energy, indicating a favorable synergy between quinoa, DSIC, and DT flour.

Table 2. Proximate composition of high-protein cookies (PC) enriched with defatted sacha inchi cake (DSIC) and defatted tarwi (DT).

Samples	Protein (%)	Moisture (%)	Lipids (%)	Ash (%)	Carbohydrates (%)	Energy (kcal)
Control	10.26 ± 0.26	10.82 ± 0.16	5.25 ± 0.35	1.51 ± 0.06	72.17 ± 0.83	377 ± 40
PC1	12.13 ± 0.29	10.67 ± 0.15	8.03 ± 0.24	1.69 ± 0.03	67.48 ± 0.71	391 ± 26
PC2	15.31 ± 0.22	10.42 ± 0.10	10.42 ± 0.28	1.89 ± 0.04	61.96 ± 0.64	403 ± 21
PC3	14.15 ± 0.11	10.78 ± 0.11	6.76 ± 0.22	1.60 ± 0.01	66.81 ± 0.45	385 ± 18
PC4	20.30 ± 0.13	10.21 ± 0.13	8.33 ± 0.26	1.54 ± 0.02	59.73 ± 0.54	395 ± 19
PC5	15.51 ± 0.15	10.55 ± 0.08	7.65 ± 0.32	1.64 ± 0.05	64.65 ± 0.60	390 ± 24

Results are expressed as means ± SD (n = 3).

3.3. Amino Acid Profile

Table 3 shows the amino acid profile of high-protein cookies.

All cookies met 9 out of 10 of the essential amino acid daily requirements. The control sample had higher levels of isoleucine (36.46 ± 1.35 mg/g protein), leucine (96.12 ± 3.63 mg/g protein), methionine + cysteine (26.36 ± 0.87 mg/g protein), phenylalanine + tyrosine (83.59 ± 4.42 mg/g protein), and valine (44.86 ± 1.71 mg/g protein) compared to the samples with DSIC and DT. However, these cookies had significantly higher amounts of tryptophan (ranging from 9.45 ± 0.45 mg/g protein to 17.79 mg/g protein) and threonine (ranging from 40.89 ± 1.51 mg/g protein to 45.11 ± 1.96 mg/g protein, except for sample PC4). Additionally, PC5 demonstrated a more balanced proportion of essential amino acids due to the blend of quinoa, DSIC, and DT, suggesting a beneficial synergy and complementarity among the three ingredients. Nonetheless, lysine was deficient across all samples (ranging from 30.34 ± 1.09 mg/g protein to 34.48 ± 0.64 mg/g protein).

Table 3. Amino acid profile (AAP) (mg/g protein) of high-protein cookies (PC) enriched with defatted sacha inchi cake (DSIC) and defatted tarwi (DT).

AAP	Control	PC1	PC2	PC3	PC4	PC5	FAO [a]
Histidine	21.08 ± 0.35	19.04 ± 1.42	18.61 ± 0.48	21.28 ± 1.20	20.81 ± 1.37	21.50 ± 2.05	18
Isoleucine	36.46 ± 1.35	34.41 ± 0.96	34.09 ± 1.54	31.83 ± 0.91	31.21 ± 0.96	35.18 ± 1.42	30
Leucine	96.12 ± 3.63	92.63 ± 4.27	88.05 ± 3.58	87.39 ± 3.14	78.76 ± 2.40	92.44 ± 1.28	59
Lysine	33.76 ± 2.79	31.60 ± 0.45	30.34 ± 1.09	31.94 ± 0.75	34.48 ± 0.64	31.45 ± 0.17	45
Methionine + Cysteine	26.36 ± 0.87	24.19 ± 3.17	23.45 ± 1.15	17.75 ± 2.14	16.43 ± 1.22	22.31 ± 1.86	22
Phenylalanine + Tyrosine	83.59 ± 4.42	78.15 ± 2.34	74.15 ± 2.98	79.63 ± 1.66	76.60 ± 2.86	77.22 ± 2.70	38
Threonine	37.39 ± 0.60	41.05 ± 1.15	45.11 ± 1.96	38.46 ± 1.13	35.99 ± 1.57	40.89 ± 1.51	23
Tryptophan	8.75 ± 0.10	13.58 ± 0.77	17.79 ± 1.63	9.45 ± 0.45	13.15 ± 0.19	12.50 ± 1.17	6
Valine	44.86 ± 1.71	42.79 ± 1.37	41.95 ± 1.82	38.01 ± 0.96	34.88 ± 1.32	41.14 ± 1.43	39

Results are expressed as means ± SD (n = 3). [a] Food and Agriculture Organization of the United Nations.

3.4. TPC, Antioxidant Activity, In Vitro Protein Digestibility (IVPD), and Oxidative Stability

The results for antioxidant activity, in vitro protein digestibility, and oxidative stability are presented in Table 4.

Table 4. Total phenolic content (TPC), antioxidant activity by DPPH, in vitro digestibility, and acid index of high-protein cookies (PC) enriched with defatted sacha inchi cake (DSIC) and defatted tarwi (DT).

Samples	TPC (μg GAE/g Cookie)	DPPH (μg trolox/g Cookie)	In Vitro Digestibility (%)	Acid Index
Control	415 ± 21	1505 ± 64	73.81 ± 0.69	0.23 ± 0.03
PC1	368 ± 23	1399 ± 35	71.63 ± 0.58	0.48 ± 0.03
PC2	346 ± 15	1286 ± 62	70.91 ± 0.45	0.66 ± 0.03
PC3	435 ± 15	1637 ± 41	72.15 ± 0.55	0.28 ± 0.01
PC4	469 ± 27	1755 ± 77	74.17 ± 0.49	0.22 ± 0.01
PC5	428 ± 24	1573 ± 73	72.99 ± 0.53	0.49 ± 0.01

Results are expressed as means ± SD (n = 3).

The cookies enriched with DSIC exhibited a higher total phenolic content (TPC), ranging from 435 ± 15 to 469 ± 27 μg GAE/g cookie, compared to the control sample and DT. Similarly, antioxidant activity measured using DPPH ranged from 1637 ± 41 to 1755 ± 77, which was also higher than in the control sample and DT. Furthermore, all cookies demonstrated a high in vitro protein digestibility (IPVD), ranging from 70.91 ± 0.45% to 74.17 ± 0.49%, though cookies containing DSIC had a lower IPVD due to their higher fiber content as suggested by Ruttarattanamongkol et al. [3]. The shelf life by the Rancimat® method determined that all cookies were safe for consumption for a period of around 278 h at 25 °C. Additionally, all cookies had a low acid index, ranging from 0.22 ± 0.01% to 0.66 ± 0.03% lactic acid, meeting the requirements of the Peruvian Standard NTP-206.001.

3.5. Sensory Analysis

The results of the sensory analysis are shown in Figure 1.

Overall, appearance (6.58 ± 0.28 to 6.85 ± 0.33), color (6.55 ± 0.35 to 6.98 ± 0.25), and smell (5.98 ± 0.37 to 6.57 ± 0.31) were the least variable parameters across all cookies. Although all cookies had similar moisture content, the texture of the control sample was the least favored (4.38 ± 0.28). In contrast, PC4 had a better texture (5.68 ± 0.36) and was described as having more "crunchiness" by the panelists. Flavor exhibited the greatest variance among all parameters, with PC2 scoring 3.93 ± 0.15 and PC3 scoring 6.10 ± 0.19. In general, the cookies had an overall acceptability rate of 72.3%, considering values equal to or greater than five [6,7]. The most acceptable cookie was PC3, with a score of 6.23 ± 0.35.

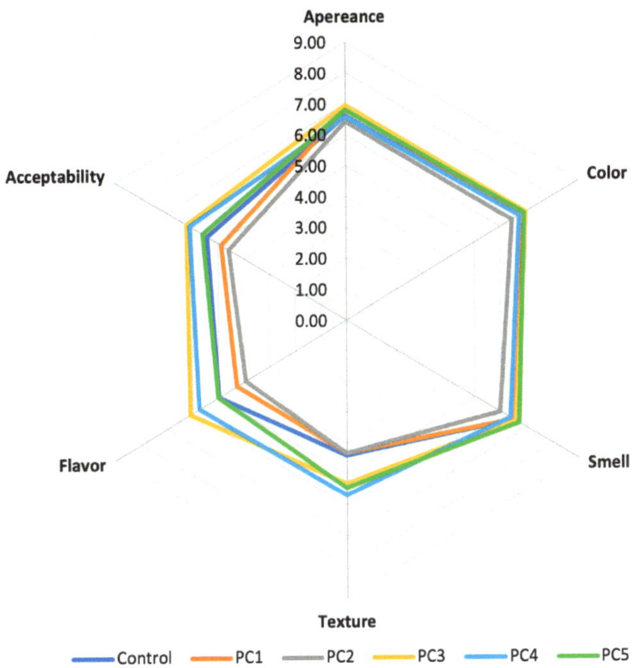

Figure 1. Sensory analysis of high-protein cookies enriched with defatted sacha inchi cake (DSIC) and defatted tarwi (DT).

4. Conclusions

In conclusion, the development of high-protein cookies enriched with defatted sacha inchi cake (DSIC) and tarwi (DT) offered a promising solution to combat child malnutrition in Peru. These enriched cookies demonstrated significantly higher protein, lipid, and antioxidant contents, providing essential nutrients that enhanced their health benefits. Sensory analysis revealed that cookies enriched with DT were particularly favored for their flavor and texture, making them more appealing to children. The notable sensory and nutritional qualities of these cookies emphasized their potential as a functional food product that could inform and support nutritional programs and policies in Peru addressing child malnutrition. By incorporating these cookies into children's diets, there will be a significant chance to improve their nutritional intake and reduce the risks associated with non-communicable diseases.

Author Contributions: All authors have contributed equally to this manuscript. Conceptualization, M.T.; methodology, M.T., N.D., A.S., R.A. and N.T.; investigation and data analysis, M.T., N.D., A.S., R.A. and N.T.; writing—original draft preparation, M.T. and N.D.; writing—review and editing, M.T. All authors have read and agreed to the published version of the manuscript.

Funding: This research received no external funding.

Institutional Review Board Statement: Not applicable.

Informed Consent Statement: Not applicable.

Data Availability Statement: Data are contained within the manuscript.

Acknowledgments: This work was developed at Laboratorio de Alimentos Funcionales of Carrera de Ingeniería Industrial and supported by Universidad de Lima, Peru.

Conflicts of Interest: The authors declare no conflicts of interest.

References

1. Pradeilles, R.; Landais, E.; Pareja, R.; Eymard-Duvernay, S.; Markey, O.; Holdsworth, M.; Rousham, E.K.; Creed-Kanashiro, H.M. Exploring the magnitude and drivers of the double burden of malnutrition at maternal and dyad levels in peri-urban Peru: A cross-sectional study of low-income mothers, infants and young children. *Matern. Child. Nutr.* **2023**, *19*, e13549. [CrossRef] [PubMed]
2. Chasquibol, N.; Alarcón, R.; Gonzales, B.F.; Sotelo, A.; Landoni, L.; Gallardo, G.; García, B.; Pérez-Camino, M.C. Design of Functional Powdered Beverages Containing Co-Microcapsules of Sacha Inchi *P. huayllabambana* Oil and Antioxidant Extracts of Camu Camu and Mango Skins. *Antioxidants* **2022**, *11*, 1420. [CrossRef] [PubMed]
3. Ruttarattanamongkol, K.; Jiapong, S. Development of direct expanded high protein snack products fortified with sacha inchi seed meal. *J. Microbiol. Biotech. Food Sci.* **2021**, *10*, 680–684. [CrossRef]
4. Carhuallanqui, S.; Ccora, A.M.; Vilcapoma, L.; Casas, J.P. Evaluación nutricional y fisicoquímica del Tarwi (*Lupinus mutabilis*) desamargado, deshidratado, tostado y envasado al vacío para consumo directo. *Prospect. Univ. Cienc. Agrar.* **2022**, *3*, 11–17.
5. Sławińka, A.; Jabłońska-Ryś, E.; Gustaw, W. Physico-Chemical, Sensory, and Nutritional Properties of Shortbread Cookies Enriched with Agaricus bisporus and Pleurotus ostreatus Powders. *Appl. Sci.* **2024**, *14*, 1938. [CrossRef]
6. Koh, W.Y.; Lim, X.X.; Tan, T.C.; Mamat, H.; Kobun, R.; Rasti, B. Utilising Spent Tea Leaves Powder as Functional Ingredient to Enhance the Quality of Non-Gluten Shortbread Cookies. *Foods* **2023**, *12*, 1557. [CrossRef] [PubMed]
7. Plustea, L.; Dossa, S.; Dragomir, C.; Cocan, I.; Negrea, M.; Obistioiu, D.; Poiana, M.A.; Voica, D.; Berbecea, A.; Alexa, E. Comparative Study of the Nutritional, Phytochemical, Sensory Characteristics and Glycemic Response of Cookies Enriched with Lupin Sprout Flour and Lupin Green Sprout. *Foods* **2024**, *13*, 656. [CrossRef] [PubMed]
8. Tapia, M.; Marimón, S.J.; Salazar, N. Development of Extruded Snacks with Protein Hydrolysed from Jumbo Squid (*Dosidicus gigas*) by-Product and Cañihua (*Chenopodium pallidicaule Aellen*). *Biol. Life Sci. Forum* **2023**, *25*, 4. [CrossRef]
9. Chasquibol, N.; Gonzales, B.F.; Alarcón, R.; Sotelo, A.; Márquez-López, J.C.; Rodríguez-Martin, N.M.; del Carmen Millán-Linares, M.; Millán, F.; Pedroche, J. Optimisation and Characterisation of the Protein Hydrolysate of Scallops (*Argopecten purpuratus*) Visceral By-Products. *Foods* **2023**, *12*, 2003. [CrossRef] [PubMed]

biology and life sciences forum

MDPI

Proceeding Paper

Morphological and Physicochemical Characterization of Native Beans Reintroduced to the Andean Zone of Jujuy, Argentina †

Cristina N. Segundo *, Silvina P. Meriles, Maria A. Giménez, Manuel O. Lobo and Norma C. Samman

Facultad de Ingeniería—Centro Interdisciplinario de Investigaciones en Tecnologías y Desarrollo Social para el NOA (CIITED), CONICET-UNJu, Ítalo Palanca 10, San Salvador de Jujuy 4600, Argentina; silvinameriles@agro.unc.edu.ar (S.P.M.); malejandragimenez@gmail.com (M.A.G.); mlobo958@gmail.com (M.O.L.); normasamman@gmail.com (N.C.S.)

* Correspondence: segundocristina@gmail.com

† Presented at the VI International Congress la ValSe-Food, Lima, Peru, 23–25 September 2024.

Abstract: The objective of this work was to characterize the morphology and physicochemical properties of 15 genotypes of native beans from the province of Jujuy, Argentina, 10 of which are *ñuñas*. The morphological descriptors used were length, width, thickness, and color. Hydration capacity (HC), popping yield (PY), proximal composition, and 100-seed weight (100 W) were also determined. The *ñuñas* presented rounded shapes and, in general, were smaller than beans since lower values of length and 100 W were observed (9–12 mm and 31.4–48.8 g, respectively, versus 13–15 mm and 40–55 g for beans). No differences were observed between both groups in width (8–7.5 mm), while thickness was more variable in *ñuñas* (5.8–7.3 mm versus 5.8–6.7 mm for beans). The *ñuñas* ranged in colors, including whitish, brown, purple, and reddish examples, with a mottled, rhomboid bicolor, and tricolor patterns. The beans ranged from light brown to dark purple, either single-colored or with wide bicolored stripes. Darker colors might indicate the presence of polyphenols and anthocyanin. The physicochemical properties depended on the genotype; the *ñuñas* presented higher HC (50–67%) and PY (20–36%). The protein content—a key characteristic of legumes—varied between 18 and 25% for all the varieties studied, while lipids ranged from 0.23 to 1.29%. In conclusion, these different characteristics of each genotype could exhibit varying behaviors in response to treatments applied for industrialization. In the canning industry, high values of HC are preferred, while PY describes the ability of *ñuñas* to expand when exposed to heat.

Keywords: beans; hydration properties; morphological descriptors; *ñuñas*; roasting

1. Introduction

The common bean, native to Central and South America, is found in northern Argentina, Bolivia, Peru, and Central America. Less well-known are *ñuñas*, which existed before the Incas [1]. Andean legumes, such as beans, *ñuñas*, and faba, are being reintroduced in Jujuy, Argentina, by the Institute of Research and Technological Development for Family Agriculture in the NOA Region (INTA-IPAF NOA) to reinforce their presence in local cuisine. There is limited knowledge about the physical characteristics of these seeds, which are important as they determine their suitability for industrial processing [1].

Andean beans have various colors [2], and determining their nutritional contribution and technological aptitudes would promote their consumption. Their proteins have high lysine and minerals [3]. Flours from commercial beans grown in northern Argentina are an important source of dietary fiber, proteins, phytosterols, and γ-tocopherol. Additionally, flours obtained from pigmented beans contain phenolic compounds and higher antioxidant activities [3] than those from non-pigmented ones. The aim was to characterize the morphology and physicochemical properties of 15 genotypes of native beans from Jujuy, 10 of which are *ñuñas*. Revaluing their properties would represent an alternative for commercialization, primarily due to their potential use in the production of healthy and gluten-free foods and would also contribute to the conservation of the biodiversity of these crops, giving rise to different uses and adding value to local cuisine.

2. Materials and Methods

The seeds were provided by INTA-IPAF NOA. Its production was carried out in its experimental field (2190 masl), Maimará, Jujuy, for two consecutive years (2023 and 2024). They were numbered from 1 to 15. Samples 2, 3, 4,5,7, 8, 9, 12, 13, and 15 are *ñuñas*, and the rest are beans (see Figure 1).

Figure 1. Andean beans from Jujuy, Argentina, 10 of which are *ñuñas* (num. 2, 3, 4, 5, 7, 8, 9, 12, 13). Images of native beans and beans after thermal treatment and hydration. N: native; R: roasted; H: hydrated.

Morphological characterization was conducted according to descriptors for the common bean (*Phaseolus vulgaris* L.) proposed by the International Board of Plant Genetic Resources (IBPGR, 1982; now, IPGRI) [4], detailed in Table 1.

Fresh beans (20 g) were roasted using an electric oven (Ken Brown, Ciudad, País) at 250 °C for 10 min. Time of first popping (T) was recorded. Popped beans were counted, and the volume of five expanded beans (VE) was recorded via displacement of ethanol. Popping yield (PY) was determined according to Equation (1):

$$PY\ (\%) = \frac{Wfpg + Wspg}{W} \times 100, \tag{1}$$

where Wpg denotes fully popped grains weight, $Wspg$ denotes semi-popped grains weight, and W denotes grains before popping total weight.

Hydration capacity (HC) and swelling capacity (SC) were expressed as the amount of water that seeds absorbed after overnight soaking in excess of water. HC and SC are expressed as percentages in weight and volume, respectively (Equations (2) and (3)):

$$C\ (\%) = \frac{W2 - W1}{W1} \times 100, \tag{2}$$

where W_1 is the weight of seeds before soaking, and W_2 is weight of seeds after soaking;

$$SC\ (\%) = \frac{V2 - V1}{V1} \times 100, \tag{3}$$

where V1 is volume of seeds before soaking and V2 is volume of seeds after soaking.

Table 1. Bean descriptors used in the present work *.

Shape	**1: Round; 2: Oval; 3: Cuboid; 4: Kidney Shape; 5: Truncate Fastigate**
	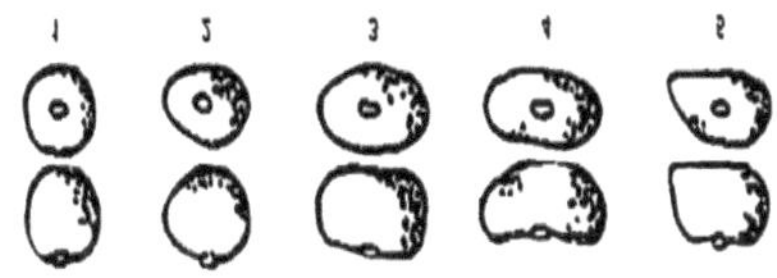
Seed coat pattern	0: absent; 1: constant mottled; 2: striped; 3: rhomboid spotted; 4: speckled; 5: circular mottling; 6: marginal color pattern; 7: broad stripped; 8: bicolor; 9: spotted bicolor; 10: pattern around hilum; 11: other (tricolor) 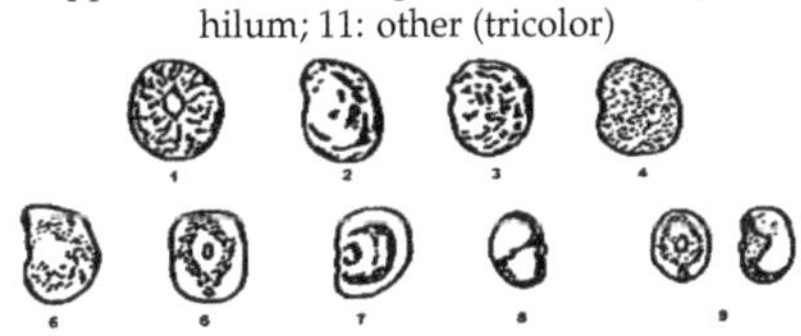
Ground color	1: black; 2: dark brown; 3: brown; 4: grey, brownish to greenish; 5: pure white; 6: whitish; 7: dark pink; 8: dark purple; 9: cream; 10: dark reddish brown
Brilliance	1: matt; 2: medium; 3: shiny
Secondary and Tertiary ground color	See ground color codes
Seed dimensions	Average of 10 seeds (mm): length (measured parallel to the hilum), width, and height (measured from hilum to opposite side)
Seed weight	Weight of 100 seeds (g)

* Adapted from IBPGR: International Board Plant Genetic Resources, 1982.

Whole beans were finely ground, and proximate chemical composition was determined according to AOAC official methods [5]. All measurements and treatments were conducted in triplicate. Data were analyzed using XLSTAT software (V2008.1.50162).

3. Results and Discussion

3.1. Evaluation of Morphological and Technological Quality

The results show that *ñuñas* (9.73–12.51 mm) are shorter compared to beans (11.06–15.27 mm), and there were no significant differences in width between beans (7.84–8.95 mm) and *ñuñas* (7.53–8.93 mm). The variability of thickness was greater among *ñuñas* than beans (Table 2).

Table 2. Proximate composition, morphological characteristics, and descriptors of Andean beans.

N°	Moisture (%)	Protein (%)	Lípid (%)	Ashe (%)	CH [b] (%)	100W [c] (g)	Length (mm)	Width (mm)	Height (mm)	Shape [d]	Brilliance [d]	Pattern [d]	Ground Color [d]	Second Ground Color [d]	Tertiary Ground Color [d]
1	10.95 ± 0.14 ^{f}	19.99 ± 0.92 ^{de}	0.68 ± 0.02 ^{bc}	5.18 ± 0.01 ^{f}	63.28	54.49 ± 0.60 ^{ef}	14.81 ± 0.15 ^{d}	8.80 ± 0.81 ^{b}	6.01 ± 0.48 ^{a}	4	2	7	3	1	0
2	11.60 ± 0.03 ^{gh}	25.07 ± 1.13 ^{g}	0.35 ± 0.02 ^{a}	4.53 ± 0.27 ^{de}	58.45	43.482 ± 0.43 ^{c}	12.51 ± 0.87 ^{c}	8.23 ± 0.60 ^{ab}	6.10 ± 0.60 ^{a}	1	2	0	1	0	0
3	10.31 ± 0.04 ^{d}	16.73 ± 0.14 ^{b}	1.10 ± 0.09 ^{g}	4.00 ± 0.03 ^{abc}	67.84	48.79 ± 0.64 ^{de}	11.92 ± 1.23 ^{bc}	8.93 ± 0.47 ^{b}	7.37 ± 0.56 ^{c}	1	1	8	14	9	0
4	11.76 ± 0.04 ^{h}	16.86 ± 0.27 ^{b}	0.99 ± 0.05 ^{ef}	4.36 ± 0.32 ^{cde}	66.02	40.83 ± 0.74 ^{c}	10.15 ± 0.51 ^{b}	8.03 ± 0.32 ^{a}	7.26 ± 0.31 ^{c}	1	2	1	2	3	0
5	10.83 ± 0.07 ^{f}	17.68 ± 0.36 ^{bc}	0.57 ± 0.00 ^{b}	4.70 ± 0.30 ^{ef}	66.22	31.37 ± 0.24 ^{a}	11.73 ± 0.62 ^{c}	7.53 ± 0.47 ^{a}	5.963 ± 0.35 ^{a}	2	1	0	8	0	0
6	10.31 ± 0.01 ^{d}	18.15 ± 1.2 ^{ef}	0.23 ± 0.00 ^{a}	4.66 ± 0.30 ^{e}	66.65	40.11 ± 1.02 ^{bc}	13.25 ± 0.99 ^{cd}	7.84 ± 0.34 ^{a}	6.07 ± 0.34 ^{ab}	1	2	0	9	2	0
7	11.48 ± 0.06 ^{g}	17.69 ± 0.19 ^{bc}	1.26 ± 0.04 ^{g}	4.55 ± 0.31 ^{de}	65.01	39.15 ± 0.31 ^{b}	10.52 ± 0.41 ^{ab}	8.81 ± 0.39 ^{b}	7.26 ± 0.52 ^{c}	1	1	1	8	6	0
8	10.88 ± 0.09 ^{f}	16.73 ± 0.01 ^{b}	1.18 ± 0.17 ^{g}	3.62 ± 0.07 ^{a}	67.58	39.85 ± 1.20 ^{b}	10.53 ± 0.37 ^{ab}	8.32 ± 0.55 ^{ab}	7.26 ± 0.25 ^{c}	1	2	11	3	6	2
9	11.44 ± 0.05 ^{g}	16.97 ± 0.12 ^{b}	1.09 ± 0.10 ^{g}	3.84 ± 0.03 ^{ab}	66.65	35.42 ± 1.18 ^{ab}	10.22 ± 0.50 ^{a}	8.15 ± 0.46 ^{ab}	6.56 ± 0.54 ^{ab}	1	1	4	9	2	0
10	10.64 ± 0.09 ^{e}	17.09 ± 0.04 ^{b}	1.63 ± 0.01 ^{h}	4.30 ± 0.03 ^{bcde}	66.32	55.28 ± 1.18 ^{f}	14.40 ± 0.85 ^{d}	8.96 ± 0.47 ^{b}	6.72 ± 0.38 ^{b}	2	2	0	10	0	0
11	10.78 ± 0.18 ^{ef}	17.22 ± 0.22 ^{bc}	0.78 ± 0.03 ^{cd}	4.15 ± 0.03 ^{bcd}	67.07	52.76 ± 0.96 ^{e}	15.27 ± 1.04 ^{e}	8.86 ± 0.42 ^{ab}	5.89 ± 0.32 ^{a}	4	2	0	3	0	0
12	10.08 ± 0.04 ^{c}	17.48 ± 0.89 ^{b}	0.95 ± 0.02 ^{e}	3.97 ± 0.05 ^{abc}	67.51	33.54 ± 0.10 ^{a}	9.831 ± 0.49 ^{a}	8.13 ± 0.36 ^{a}	6.92 ± 0.34 ^{c}	1	2	0	4	0	0
13	12.16 ± 0.02 ^{i}	13.76 ± 0.08 ^{a}	1.14 ± 0.11 ^{fg}	4.38 ± 0.43 ^{cde}	68.54	46.05 ± 0.46 ^{d}	11.07 ± 0.96 ^{bc}	8.91 ± 0.41 ^{b}	7.50 ± 0.56 ^{b}	1	2	8	1	5	0
14	8.96 ± 0.01 ^{a}	18.73 ± 0.06 ^{cd}	1.29 ± 0.15 ^{g}	4.67 ± 0.18 ^{e}	66.35	44.58 ± 0.45 ^{d}	13.38 ± 1.50 ^{c}	8.27 ± 0.80 ^{ab}	6.16 ± 0.53 ^{ab}	2	2	2	9	2	0
15	9.15 ± 0.02 ^{b}	21.94 ± 0.25 ^{f}	0.88 ± 0.02 ^{de}	3.96 ± 0.01 ^{abc}	64.07	41.80 ± 1.63 ^{c}	9.73 ± 0.72 ^{a}	8.32 ± 0.44 ^{b}	6.57 ± 0.49 ^{bc}	1	2	1	7	9	0

[a] Mean ± SD (different letters within a column indicate significant differences ($p < 0.05$)); [b] CH: carbohydrates, calculated by difference 100–(moisture + protein + lipids + ash); [c] 100W: 100 seed weight; [d] See the codes used in each descriptor in Table 1. Letters e, f, g, h, i indicate significant differences.

Andean beans are categorized as medium and large according to the 100-seed weight [6], while *ñuñas*, on average, are smaller. Nevertheless, according to the IBPGR [4], based on seed length, Andean beans are classified as medium (1–2 cm length, approx.) and *ñuñas* as small (<1 cm length, approx.). Cruz Balarezo et al. [1] reported similar findings for native *ñuñas* from Peru and Bolivia. From a technological standpoint, these characteristics impact grinding performance, cooking time, nutritional value, and appearance.

Ñuñas are typically round, though one sample (N° 2) was classified as oval. Beans were described as round, oval, and kidney-shaped (Table 2). The brilliance was generally medium for both. The most distinctive feature of Andean beans is their pattern diversity color, overshadowing their nutritional potential. *Ñuñas* exhibited a broad range of primary colors (black, brown, grey, dark pink, dark purple, cream, and dark reddish brown). Some *ñuñas* are single-colored (samples N°. 2, 5, 12), while bicolor ones combine lighter colors (whitish, cream) or darker hues such as brown (Table 1). Only ñuña N° 8 showed three colors. Among the beans, three were bicolor, featuring dark broad stripes (black and brown). Other observed color patterns in *ñuñas* included constant mottled, rhomboid spotted, speckled, and bicolor. These color variations in the seed coat may be associated to the presence of phenolic compounds [6].

3.2. Chemical Composition

Significant differences were observed between beans and *ñuñas* ($p < 0.05$) for all macro components analyzed (Table 2). The moisture content was higher in *ñuñas* than beans. In all cases, the lipid content was less than 1.63%. Both beans and *ñuñas* exhibited high protein and carbohydrate content. Similar studies on beans [2,3] suggest high fiber content. This makes them a significant source of nutrients with functional potential. The results align with those reported by Nagai et al. [3], and also with native varieties from Peru and Bolivia [2].

3.3. Hydration Capacity

Figure 1 shows that all grains increased in size after soaking. HC and SC of *ñuñas* were significantly higher than those of beans (Figure 2). Water retention is associated with chemical composition as it depends on the retention of water molecules by the hydrophilic components of bean macromolecules [7]. Kajiwara et al. [7] suggest the formation of complexes between carbohydrates, polyphenols, and proteins during maceration, which enhances the physical retention of water but increases cooking time. This phenomenon could explain the high SC observed in *ñuña* N°4 (dark brown color).

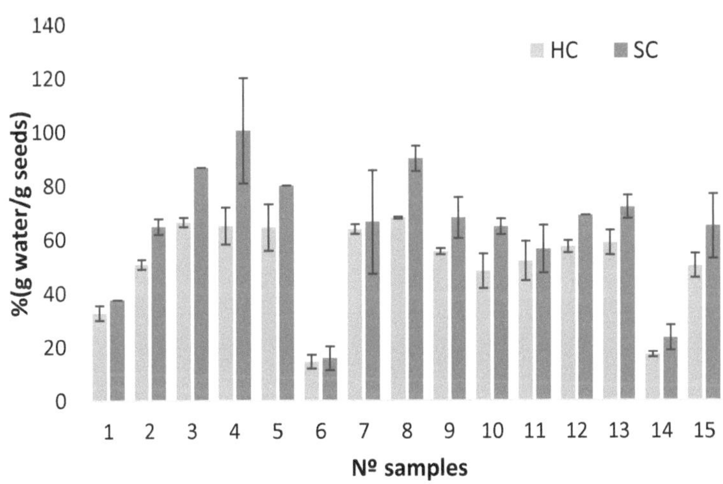

Figure 2. Hydration properties (HC: hydration capacity; SC: swelling capacity) of Andean beans.

3.4. Roasted Grain Quality

Figure 1 indicates that only *ñuñas* demonstrated expansion capacity during roasting, as evidenced by the separation of the seed coat at the center of the bean. PY, VE, and T (Table 3) showed significant differences among the analyzed *ñuñas*. Genotype N°12 exhibited the highest PY, sample N° 3 had the greatest VE, and sample N°9 required the shortest time to initiate popping (3.34 s). The popping capacity is a significant quality in *ñuña*-type beans, which is influenced by genetic variability as well as non-genetic factors such as seed shape, its inelastic coat, and the quantity and quality of stored starch [1].

Table 3. Roasting and popping properties of *ñuñas*.

N°	Popping Yield (%)	Vol. 5 Popped Beans (mL)	Time First Popping (s)
2	20.5 ± 1.2 ^{cd}	3.87 ± 0.17 ^{bc}	6.46 ± 0.04 ^{f}
3	15.9 ± 0.1 ^{b}	4.10 ± 0.14 ^{c}	4.83 ± 0.16 ^{c}
4	24.8 ± 0.04 ^{e}	2.70 ± 1.13 ^{b}	4.44 ± 0.04 ^{b}
5	18.3 ± 1.24 ^{c}	3.50 ± 0.70 ^{bc}	3.7 ± 0.16 ^{a}
7	21.3 ± 0.01 ^{d}	3.00 ± 0 ^{b}	4.78 ± 0.59 ^{d}
8	36.0 ± 1.35 ^{f}	3.25 ± 0.35 ^{b}	5.22 ± 0.38 ^{d}
9	14.9 ± 1.40 ^{b}	2.50 ± 0.70 ^{b}	3.34 ± 0.49 ^{a}
12	25.2 ± 1.55 ^{e}	2.25 ± 0.35 ^{a}	4.18 ± 0.91 ^{a}
13	23.6 ± 2.96 ^{de}	4.00 ± 0 ^{bc}	5.22 ± 0.31 ^{d}
15	7.9 ± 1.22 ^{a}	1.05 ± 0.35 ^{a}	5.26 ± 0.30 ^{d}

Letters a–f indicate significant differences.

4. Conclusions

The native legumes studied demonstrate significant potential as sources of proteins, minerals, and possibly functional compounds. The native genotypes were classified as medium or large based on the weight of 100 grains, or as small or medium based on their length. The presence of dark colors in the seed coat of some genotypes is likely associated with antioxidant compounds. The hydration properties are particularly significant, especially in certain *ñuña* genotypes. The *ñuñas* are proposed as popping beans due to their ability to expand when exposed to heat and could produce a snack-type food with high protein content following a short thermal treatment (3–7 min). Revaluing Andean bean properties would represent an alternative for commercialization and would contribute to the conservation of the biodiversity of these crops, fostering different uses and adding value to local cuisine.

Author Contributions: Conceptualization, C.N.S. and S.P.M.; methodology, C.N.S.; software, S.P.M.; validation, C.N.S., S.P.M. and M.A.G.; formal analysis, S.P.M.; investigation, C.N.S.; resources, C.N.S. and S.P.M.; data curation, S.P.M.; writing—original draft preparation, C.N.S. and S.P.M.; writing—review and editing, C.N.S., S.P.M., M.A.G., M.O.L. and N.C.S.; visualization, S.P.M.; supervision, N.C.S.; project administration, M.O.L.; funding acquisition, M.O.L. All authors have read and agreed to the published version of the manuscript.

Funding: This research received no external funding.

Institutional Review Board Statement: Not applicable.

Informed Consent Statement: Not applicable.

Data Availability Statement: Data are contained within the article.

Acknowledgments: The authors express their thanks to INTA- IPAF NOA for providing the seeds as part of their recovery program concerning the recovery and reintroduction processes of Andean beans in the region. They also thank the Ia ValSe-Food-CYTED Network for its support and to Universidad de Lima.

Conflicts of Interest: The authors declare no conflict of interest.

References

1. Cruz Balarezo, J.; Camarena Mayta, F.; Baudoin, J.P.; Huaringa Joaquín, A.; Blas Sevillano, R. Evaluación agromorfológica y caracterización molecular de la ñuña (*Phaseolus vulgaris* L.). *Idesia (Arica)* **2009**, *27*, 29–40. [CrossRef]
2. Kajiwara, V.; ModaCirino, V.; dos Santos Scholz, M.B. Studies on nutritional and functional properties of various genotypes of Andean beans. *J. Food Sci. Technol.* **2021**, *59*, 1468–1477. [CrossRef] [PubMed]
3. Nagai, N.F.; Argel, N.S.; Andrés, S.C. Physicochemical characterization and functional properties of flours from North-western Argentina bean (*Phaseolus vulgaris* L.) cultivars. *Cereal. Chem.* **2022**, *99*, 850–859. [CrossRef]
4. International Board Plant Genetic Resources (IBPGR). *Descriptors for Phaseolus vulgaris* L.; Food and Agiculture Organization of de United Nations: Roma, Italy, 1982; 32p.
5. AOAC. *Official Methods of Analysis of AOAC International*, 17th ed.; Association of Official Analytical Chemists (AOAC) International: Gaithersburg, MD, USA, 2000.
6. Jamanca Gonzales, N.C.; Ocrospoma Dueñas, R.W.; Eguilas Caushi, Y.M.; Padilla Fabian, R.A.; Silva-Paz, R.J. Food grain quality: Analysis of physical, biometric, and colorimetric properties to promote consumption. *Heliyon* **2024**, *10*, e29234. [CrossRef] [PubMed]
7. Kajiwara, V.; Moda-Cirino, V.; dos Santos Scholz, M.B. The influence of chemical composition diversity in the cooking quality of Andean bean genotypes. *Food Chem.* **2021**, *339*, 127917. [CrossRef] [PubMed]

biology and life sciences forum

MDPI

Proceeding Paper

The Effect of the Incorporation of Dried Moringa Leaf Powder on the Physicochemical and Sensory Properties of Snack Crackers †

Lourdes Cervera-Chiner [1], Andrea Sanz [1], Francisco José García-Mares [2], María Luisa Castelló [1] and María Dolores Ortolá [1,*]

1 Food Engineering Research Institute—FoodUPV, Universitat Politècnica de València, Camino de Vera s/n, 46022 Valencia, Spain; loucerch@upv.edu.es (L.C.-C.); ansante@upv.edu.es (A.S.); mcasgo@upv.edu.es (M.L.C.)

2 Department of Hydraulic Engineering and Environment, Universitat Politècnica de València, Camino de Vera s/n, 46022 Valencia, Spain; fjgarcia@upv.edu.es

* Correspondence: mdortola@upv.edu.es

† Presented at the VI International Congress la ValSe-Food, Lima, Peru, 23–25 September 2024.

Abstract: Food reformulation has become a critical concern for the food industry due to society's growing interest in improving nutritional profiles. In this context, *Moringa oleifera*, a plant native to India with high nutritional value, offers an alternative for enriching food products. Its abundant antioxidants, proteins and fibers make it an attractive choice. This study aimed to assess the impact of substituting wheat flour with dried moringa leaf powder in snack crackers. These were prepared using 53% (w/w) wheat flour and substituting part of it with different replacement percentages (1, 2.5, 5, 7.5 and 10% (w/w)) of dried moringa leaf powder. The baked snacks were characterized in terms of moisture, aw, optical properties, mechanical properties, antioxidant capacity, total phenol content, protein content and energy value. In addition, a sensory analysis was carried out to evaluate the acceptability of the crackers. The results indicated that cracker thickness and volume remained constant across all formulations. As moringa incorporation increased, weight loss decreased. The high water-holding capacity of moringa leaf powder and its protein content contributed to keep the same moisture content and reduce water activity in the crackers, resulting in decreased firmness. The snacks exhibited a greener color with brownish tones as moringa replacement levels rose. Antioxidant capacity (up to 251 ± 13 mg Trolox E/100 g snack) and total phenol content (up to 1172 ± 288 mg Galic acid/100 g snack) were higher with greater moringa inclusion, remaining stable after baking. The protein content increased, allowing all crackers to be labeled as a "protein source" since the energy value due to protein was higher than 12%. However, judges found the color, aroma and flavor attributes of the highest moringa content (10%) crackers too intense. In conclusion, replacing up to 5% of wheat flour with dried moringa leaf powder in snack crackers could enhance their nutritional profile while maintaining consumer acceptance.

Keywords: crackers; formulation; moringa; sensory; snacks; texture

1. Introduction

Spain is one of the main snack-consuming countries in Europe; 8 out of 10 Spaniards consume snacks on a regular basis, either weekly or daily, with an expenditure that reaches EUR 57 per capita per year, only behind the United Kingdom and the Netherlands [1]. In 2022, the snacking sector had a turnover of EUR 2316 million, experiencing 10% growth in value. Snack crackers are the most widely consumed snack. However, the sector faces the difficulty of combining the tasty with the nutritious, as well as the priority use of natural ingredients. Thus, the inclusion of components obtained from fruits and vegetables, plants or seeds is becoming more and more common. In this context, *Moringa oleifera*, a plant native to India (although it has adapted to other tropical and subtropical areas around the world), little known in Europe, but widely cultivated in Africa and Central and South America, is a crop that has low water requirements and can withstand high temperatures and long periods of drought [2]. Moreover, moringa has a high nutritional value; its nutritional composition in vitamins, amino acids and micronutrients is atypical for a plant. Its leaves stand out for their high protein content (30%); it contains 18 of the 20 essential amino acids; it also has a vitamin C content seven times higher than that of oranges and a vitamin A content four times higher than that of carrots and a calcium content four times higher than milk; it also has a high content of antioxidants and fiber [3]. For all these reasons, leaves of moringa offer an attractive alternative as an ingredient for food reformulation and enriching food products.

For all of that, the objective of this study is to evaluate the effect of substituting wheat flour with dried moringa leaf powder in snack crackers at different proportions. For this purpose, its physicochemical properties (moisture, a_w, color, texture), antioxidant capacity, total phenol content, protein content and energy value have been calculated, and, finally, a sensory analysis has been carried out.

2. Materials and Methods

2.1. Moringa Powder Production

Moringa powder was obtained from moringa leaves collected from trees in the experimental plot at the Universitat Politècnica de València (Spain). The leaves were dried at 50 °C for 8 h in a convective tray oven (CLK 750 TOP+, POL-EKO, Wodzisław, Poland). Subsequently, the leaves were crushed at 10,200 rpm using a food processer (Thermomix TM31 Vorwerk, Wupertal, Germany). The powder was sieved with a sieve with a mesh size of 0:100 mm and stored in airtight glass jars in the dark at room temperature.

2.2. Snack Crackers Preparation and Formulation

The snacks were prepared using 53% (*w*/*w*) flour and substituting part of it with different replacement percentages (1, 2.5, 5, 7.5 and 10% (*w*/*w*)) of dried moringa leaf powder (Table 1). First, flour was mixed with moringa powder, sugar, salt, baking powder and melted butter in a food processer (Thermomix, TM31 Vorwerk, Wupertal, Germany) for 4 s at the speed 13,000 rpm. Then, milk was added and programmed for 6 s at the speed 4600 rpm. The dough was then left to rest at room temperature for one hour before being rolled out with a rolling pin to obtain a flat dough 2 mm thick. The dough was then cut with a pastry cutter into squares of 46 mm on each side. Finally, the cookies were baked for 15 min at 180 °C with ventilation in a multifunction oven (Model Bejublad, 403.009.02, Electrolux, Stockholm, Sweden). The samples were allowed to temper for at least two hours and stored in airtight containers until analysis.

Table 1. Percentage of ingredients for each snack cracker formulation.

Formulation	Wheat Flour	Moringa Powder	Salt	Sugar	Baking Powder	Butter	Milk
S0	53.1	0	2	0.9	1	8	35
S1	52.6	0.5	2	0.9	1	8	35
S2.5	51.7	1.4	2	0.9	1	8	35
S5	50.4	2.7	2	0.9	1	8	35
S7.5	49.1	4.0	2	0.9	1	8	35
S10	47.8	5.3	2	0.9	1	8	35

2.3. Analytical Determinations

The moisture of the snacks was analyzed in triplicate according to the gravimetric method (AOAC, 2000), by drying in an oven (J.P SELECTA, model conterm type poupinel 2000201, Barcelona, Spain) at 60 °C until a constant weight was reached. Water activity analysis was performed using a dew point hygrometer (AquaLab, Decagon Devices, Inc., model 4TE, Pullman, Washington, DC, USA) at 25 °C, in triplicate for each snack formulation.

A cutting test was performed using a cutting probe that completely fragmented the snack with a feed rate of 0.5 mm/s using a universal press (TA.XT. plus texture analyzer, Microsystems stable, Godalming, UK). The peak force value and the area under the curve were recorded. The test was performed for 12 samples of each formulation.

Color measurements were analyzed using a spectrocolorimeter (Konica Minolta, Inc., model CM-3600d, Tokyo, Japan), using the CIE L*a*b* reference system with the D65 illuminant and 10° observer. The L* luminosity and a* and b* coordinates were recorded. In addition, color differences from the control snacks (S0) were calculated using Equation (1):

$$\Delta E^* = [(\Delta L^*)^2 + (\Delta a^*)^2 + (\Delta b^*)^2]^{1/2} \quad (1)$$

The color of the upper side of 12 samples of each formulation was measured.

The antioxidant capacity of the cookies was performed by the 2,2-diphenyl-1-picrylhydrazyl free radical scavenging method. A total of 0.5 g of sample was taken from each formulation (S0, S1, S2.5, S5, S7.5 and S10) and diluted in 5 mL of 80:20 methanol (methanol–water). They were then shaken for 1 h at 300 rpm and centrifuged for 10 min at 10,000 rpm. The absorbance was measured at a wavelength of 515 nm in a spectrophotometer (Thermo Fisher Scientific, Inc. Helios Zeta UV–VIS, Waltham, MA, USA). All formulations were measured in triplicate. Results were expressed as mg Trolox equivalent/100 g snack.

The phenol total content was measured by the Folin–Ciocalteu method. For this, 0.5 g of sample was taken from each formulation and diluted in 5 mL of 80:20 methanol (methanol–water) and shaken for one hour at 300 rpm. Subsequently, they were centrifuged for 10 min at 10,000 rpm. The absorbance was measured with a spectrophotometer (Thermo Fisher Scientific, Inc. Helios Zeta UV–VIS, Waltham, MA, USA) at a wavelength of 760 nm. The results were expressed as mg gallic acid equivalent/100 g of snack.

The protein content in the snacks was determined by digestion and distillation using the Kjeldahl method. The measurements were made in triplicate. The total protein content in the sample was obtained by considering the total nitrogen content obtained multiplied by the conversion factor (6.25), which is recommended for vegetable products. The results were expressed as grams of protein/100 g of snack.

2.4. Energy Value

The energy value was calculated for each of the snack formulations, expressing the result in kilocalories per unit weight. The energy value contributed by each ingredient was obtained from the databases of the U.S. Department of Agriculture (U.S. Department of Agriculture, 2024) [4] except for the caloric value of the dry powder of *Moringa oleifera* leaves that was obtained according to Alvarez [5]. In addition, the energy value provided by the proteins in each snack formulation was calculated. This value was calculated by multiplying by 4 the protein content of each formulation, thus applying the proportionality principle where 1 g of protein is equal to 4 kcal and considering the water content of each cracker after baking.

2.5. Sensory Analysis

The potential sensory acceptability of 4 snack formulations (S0, S2.5, S5 and S10) was studied using a panel composed of 31 tasters aged between 18 and 65 years. The questionnaire was conducted using the Forms platform of Office 365. Appearance, color, aroma, texture, crunchiness and taste were evaluated using a nine-point hedonic scale [6,7], considering 1 "I dislike it very much" to 9 "I like it very much". In addition, a penalty analysis was performed to know whether an attribute located above or below the "Just about right" had penalized in the overall acceptance score.

3. Results and Discussion

After baking, no significant differences were observed in thickness (6.8 ± 1.1 mm) and the final volume of the snacks (11 ± 2 cm^3). Therefore, replacing wheat flour and, consequently, gluten with moringa powder did not decrease the dough's expansion capacity during baking. However, as the moringa powder content increased, the weight loss after baking decreased (Figure 1), which is likely related to some possible chemical reactions between moringa powder and other components of the recipe. The moisture content was similar for all cases (30 ± 3 g water/100 g crackers), and the water activity ranged between 0.68 and 0.52, with lower values observed when the replacement was higher than 5%. This phenomenon could be related to moringa leaf powder's high water retention capacity and its protein content, establishing a direct interaction between these molecules and water.

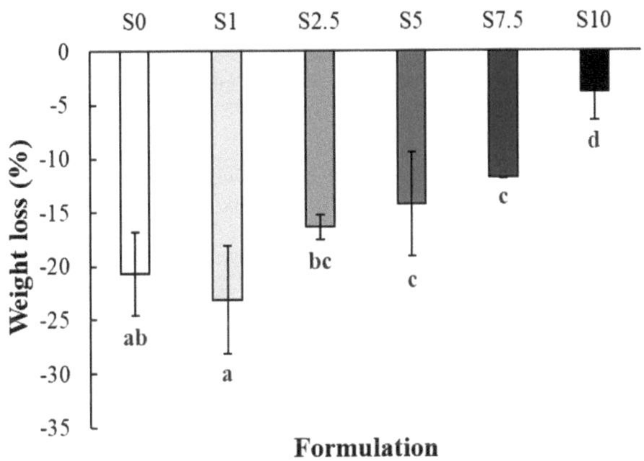

Figure 1. Weight loss of snack crackers after baking. Equal letters indicate homogeneous groups in the ANOVA (95% significance level).

The shear test results (Figure 2) showed a significant decrease in maximum strength and area under the curve from a moringa replacement level of 5%, indicating a lower hardness of the cracker formulations as the level of flour replacement with moringa leaf powder increases. These results are contrary to those reported by other authors incorporating moringa leaf powder in sweet doughs [8]. According to Alam et al. [9], the texture of cookies depends mainly on the speed of dough development and the proportion of sugar used. Therefore, the lower hardness of these cookies could also be related to the low sugar content.

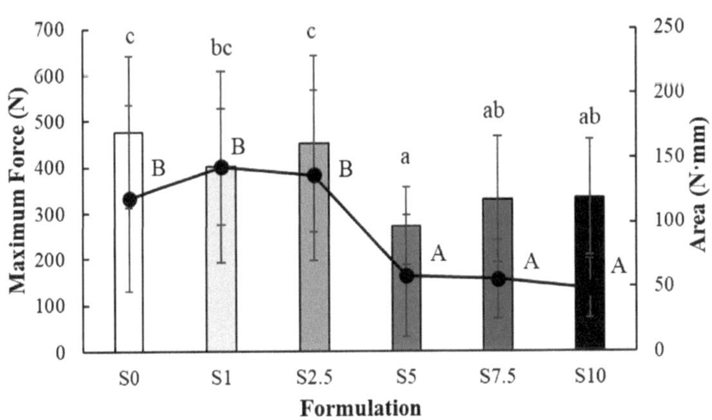

Figure 2. Maximum force (bars) and area under the curve (line) of the snack crackers shear test. Equal letters indicate homogeneous groups in the ANOVA (95% significance level), being lower case letters for maximum force and upper case letters for area.

Table 2 shows the optical properties analyzed on the surface of the different formulations, luminosity, a* and b* coordinates and color difference (ΔE) to the control snack. The replacement of wheat flour with moringa decreased the values of the a* coordinate, showing the greenish shades characteristic of this powder. In addition, significant changes were also observed in the b* coordinate related to the brownish tones. Thus, the control cracker and those with higher moringa powder content presented lower b* values. This fact would be related to the Maillard reactions that occur due to the content of both reducing sugars and amino acids provided by the ingredients used in the formulations. Furthermore, the replacement of flour with moringa powder resulted in a significant drop in brightness. Consistent with the previous results, the color difference increased with the level of replacement, exceeding in all cases the threshold of 3, and thus color differences were perceptible by the human eye.

Table 2. Luminosity (L*), chromatic coordinates (a* and b*) and color difference (ΔE) on the surface of the different snack cracker formulations.

Formulation	L*	a*	b*	DE
S0	69 ± 3^{f}	10.0 ± 1.6^{d}	32.4 ± 1.2^{c}	3.0 ± 1.7^{a}
S1	63.2 ± 0.9^{e}	1.4 ± 1.0^{a}	37.4 ± 0.9^{e}	11.7 ± 0.6^{b}
S2.5	62 ± 0.9^{d}	1.5 ± 1.0^{a}	$37{,}8 \pm 0.7^{e}$	12.7 ± 0.8^{c}
S5	49.4 ± 0.8^{c}	2.5 ± 0.8^{b}	$34{,}7 \pm 0.7^{d}$	21.3 ± 0.7^{d}
S7.5	46.7 ± 0.8^{b}	3.7 ± 0.8^{c}	31.3 ± 1.2^{b}	23.5 ± 0.7^{e}
S10	44.7 ± 0.5^{a}	3.5 ± 0.8^{c}	29.6 ± 0.7^{a}	25.5 ± 0.5^{f}

Equal letters in the same column indicate homogeneous groups (95% significance level).

Table 3 shows the antioxidant capacity and total phenol content of the studied formulations. As can be observed, in both cases, they increase as the percentage of moringa incorporated into the cracker increases. These results may be related to the content of phytochemical compounds in moringa leaves. Specifically, moringa leaves are characterized by their high content of vitamin C, iron, carotenoids and phenolic compounds such as gallic acid, kaempferol, quercetin and astragalin [10], evidencing the relationship between antioxidant capacity and phenolic compounds found in moringa leaves which also persist after baking [11]. In addition, as the level of flour replacement with moringa leaf powder increased, the protein content increased (Table 3). These results highlight one of the main characteristics of moringa leaves, their high protein content, being approximately 30% in dry powder [8,12], while in wheat flour this value is approximately 10%. Considering the proportion of the different ingredients in each of the formulations and the final moisture content of the snacks and taking as a reference their caloric values from the USDA database [4], Table 3 calculates the calories per 100 g of each of the crackers, as well as the percentage of this caloric contribution due to protein. As expected, this percentage increased as the moringa leaf powder content increased. Moreover, it should be noted that the substitution of part of the wheat flour with this powder decreases the caloric contribution of the crackers, which may be interesting from a nutritional point of view. These results would allow labeling all the crackers evaluated in this study as a "source of protein", according to Regulation No. 1924/2006 [13] since the caloric content provided by protein is higher than 12%. However, they could not be classified as "rich in protein" since this value does not exceed 20%.

Table 3. Antioxidant capacity, total phenol content, protein content, energy value and percentage of calories contributed by protein for each snack cracker formulation.

Formulation	Antioxidant Capacity (mg Trolox E/100 g Snack)	Total Phenol Content (mg Galic Acid/100 g Snack)	Protein Content (g Prot/100 g Snack)	Energy Value (kcal/100 g Snack)	% Energy Value Due to Protein
S0	33.0 ± 0.9 [a]	268 ± 31 [a]	8.7 ± 0.6 [a]	317	14.2
S1	80.5 ± 0.5 [a]	337 ± 57 [a]	9.2 ± 0.1 [ab]	295	14.4
S2.5	87 ± 3 [a]	363 ± 76 [ab]	9.36 ± 0.04 [b]	289	14.7
S5	149 ± 9 [b]	596 ± 102 [b]	9.88 ± 0.12 [c]	287	15.3
S7.5	254 ± 23 [c]	1131 ± 91 [c]	10.34 ± 0.19 [c]	282	15.8
S10	251 ± 13 [c]	1172 ± 288 [c]	11.05 ± 0.30 [d]	274	16.4

Equal letters in the same column indicate homogeneous groups (95% significance level).

Figure 3 shows the results of the sensory analysis obtained from the survey of the attributes analyzed on a hedonic scale (1–9) for the control cracker (S0) and crackers with three different levels of flour replacement with moringa leaf powder (S2.5, S5 and S10). Significant differences between samples were observed for the attributes of appearance, color, aroma and crispness, while for the attribute of texture similar scores were achieved between samples.

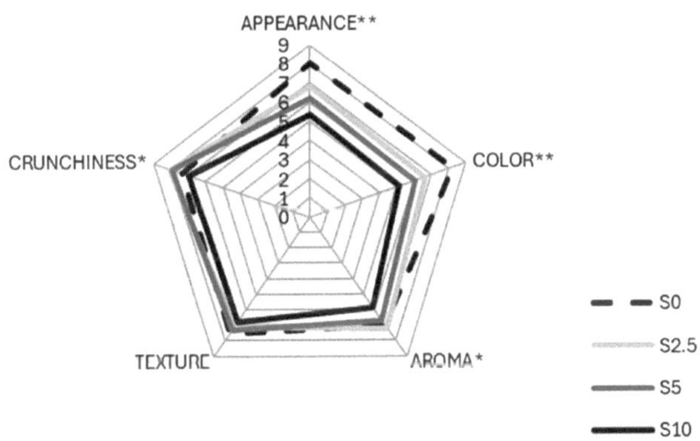

Figure 3. The results of the sensory analysis with the hedonic scale of the snack formulations studied, according to the data obtained from the ANOVA analysis; * 95% significance level and ** 99% significance level.

The penalties graph (Figure 4) relates the overall acceptance to the percentage of tasters who considered the intensity of each attribute too low or too high. Thus, S10 was penalized with a level higher than 1 and a percentage of judges higher than 20% for aroma, flavor and color because these attributes were too intense. In addition, flavor was penalized in the S5 cracker for being too intense. These results are coherent, considering that the panelists are not accustomed to the intense green color of the moringa content in the crackers, as well as their flavor and aroma. In this line, in other works, the incorporation of 10% moringa leaf powder in bakery products, specifically bread, biscuits, muffins and cookies, also had a low sensory acceptance due to the dark green color and an herbal flavor [9,14]. On the other hand, it should be noted that no parameter related to the mechanical properties of the snacks was penalized, so the differences found in the mechanical properties did not imply a rejection by the tasters.

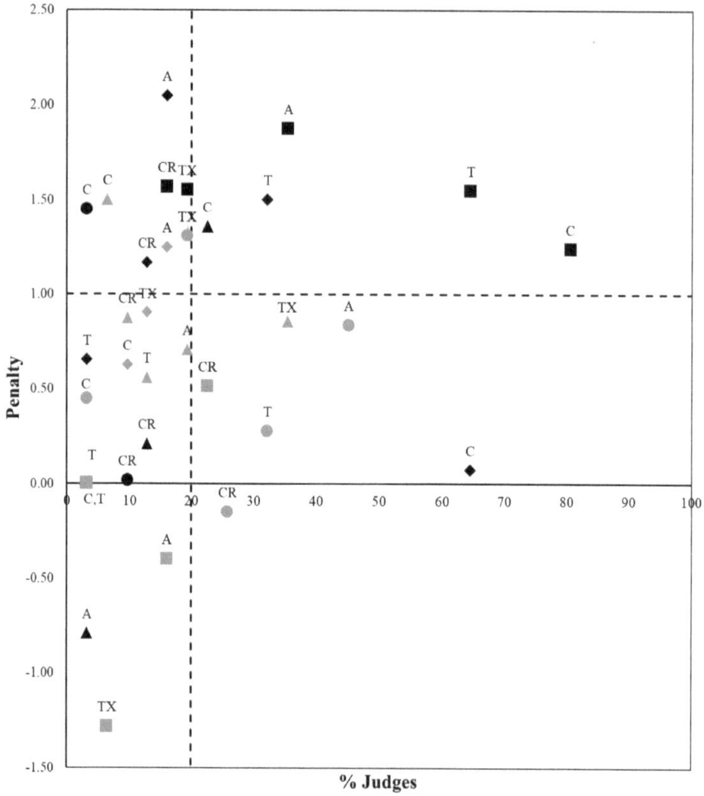

Figure 4. A graph of the penalties of the snacks evaluated in the analysis, being circle S0, triangle S2.5, rhombus S5 and square S10. Black color for too-high intensity and gray color for too-low intensity for the analyzed attributes (C: color, A: aroma, TX: texture, CR: crunchiness, T: taste).

4. Conclusions

As a general conclusion of this work, a substitution of up to 5% of wheat flour with moringa oleifera dry leaf powder in snack crackers could improve the nutritional profile of these products traditionally made only with wheat flour, with good acceptance by consumers. In addition, these crackers could be labeled, according to European legislation, as "high in protein".

Author Contributions: Conceptualization, M.D.O. and M.L.C.; methodology, M.D.O., M.L.C., F.J.G.-M. and L.C.-C.; formal analysis, L.C.-C. and A.S.; investigation, M.D.O., M.L.C., L.C.-C. and A.S.; resources, F.J.G.-M.; data curation, L.C.-C. and A.S.; writing—original draft preparation, L.C.-C. and A.S.; writing—review and editing, M.D.O., M.L.C. and L.C.-C.; supervision, M.D.O. and M.L.C.; project administration, M.D.O. and M.L.C.; funding acquisition, M.D.O., M.L.C., L.C.-C. and F.J.G.-M. All authors have read and agreed to the published version of the manuscript.

Funding: This work funded by the European Union-Next Generation EU by the "Margarita Salas" contract (2021–2024) of L. Cervera-Chiner from Universitat Politècnica de València and the Spanish Ministry of Universities with the UPV Contract and the APC was funded by Universidad de Lima (Perú).

Institutional Review Board Statement: A sensory test was conducted with human subjects. The study was conducted in accordance with the Declaration of Helsinki. The data obtained were treated in accordance with the Regulation (EU) 2016/679 of the European Parliament and of the Council of

27 April 2016, on the protection of natural persons with regard to the processing of personal data and on the free movement of such data, and repealing Directive 95/46/EC (General Data Protection Regulation).

Informed Consent Statement: Informed consent was obtained from all subjects involved in the study.

Data Availability Statement: The original contributions presented in this study are included in the article. Further inquiries can be directed to the corresponding author.

Acknowledgments: The authors thank the support provided by the Universitat Politècnica de València and La ValSe-Food-CYTED (119RT0567).

Conflicts of Interest: The authors declare no conflict of interest.

References

1. Food Retail & Service. 8 de Cada 10 Españoles Consumen Snacks de Forma Habitual. 2024. Available online: https://www.foodretail.es/food/snacks-consumo-espana-estudio-sigma_0_1804019602.html (accessed on 7 February 2024).
2. Alfaro, C.; Martínez, W. Uso Potencial de La Moringa (Moringa Oleifera, Lam) Para La Producción de Alimentos Mejorados. 2008. Available online: https://www.sica.int/documentos/uso-potencial-de-la-moringa-moringa-oleifera-lam-para-la-produccion-de-alimentos-nutricionalmente-mejorados_1_36997.html (accessed on 7 February 2024).
3. Cajamar. Moringa Oleifera: Árbol Multiusos de Interés Forestal Para El Sur de La Península Ibérica. 2016. Available online: https://www.cajamar.es/storage/documents/020-moringa-v3-1476963334-bf35c.pdf (accessed on 24 July 2024).
4. U.S. Department of Agriculture. *FoodData Central.*. Available online: https://www.usda.gov/ (accessed on 10 March 2024).
5. Alvarez, A. *Valor Nutricional de La Moringa Oleífera. Mito o Realidad;* Universidad San Francisco de Quito: San Francisco, CA, USA, 2017.
6. *UNE 87025:1996;* Análisis Sensorial. Metodología. Perfil de Textura. Asociacion Española de Normalización: Madrid, Spain, 1996.
7. *ISO 4121:2003;* Sensory Analysis—Guidelines for the Use of Quantitative Response Scales. International Organization for Standardization: Geneva, Switzerland, 2003.
8. Hedhili, A.; Lubbers, S.; Bou-Maroun, E.; Griffon, F.; Akinyemi, B.E.; Husson, F.; Valentin, D. Moringa Oleifera Supplemented Biscuits: Nutritional Values and Consumer Segmentation. *S. Afr. J. Bot.* **2021**, *138*, 406–414. [CrossRef]
9. Alam, M.A.; Alam, M.J.; Hakim, M.A.; Huq, A.O.; Moktadir, S.G. Development of Fiber Enriched Herbal Biscuits: A Preliminary Study on Sensory Evaluation and Chemical Composition. *Int. J. Nutr. Food Sci.* **2014**, *3*, 246–250. [CrossRef]
10. Barragán, L.M.; Campos Gaona, R.; Torres Castañeda, H. Fenoles Totales y Capacidad Antioxidante Del Extracto de Hojas de Moringa Oleifera En Tres Estados Fenológicos. *Acta Agron.* **2023**, *71*, 156–161. [CrossRef]
11. Famakinwa, A.; Shuttleworth, A.; Lubisi, S.; Olubi, O.; Oguntibeju, O.O.; Obilana, A. Chemical and Functional Properties of Snacks Produced from Wheat Flour Fortified with Moringa Oleifera Leaf Powder. *Afr. J. Food Agric. Nutr. Dev.* **2023**, *23*, 24467–24486. [CrossRef]
12. Olusanya, R.N.; Kolanisi, U.; van Onselen, A.; Ngobese, N.Z.; Siwela, M. Nutritional Composition and Consumer Acceptability of Moringa Oleifera Leaf Powder (MOLP)-Supplemented Mahewu. *S. Afr. J. Bot.* **2020**, *129*, 175–180. [CrossRef]
13. European Parliament and the Council of the European Union. Regulation (EC) No 1924/2006 of the European Parliament and of the Council of 20 December 2006 on Nutrition and Health Claims Made on Foods. Official Journal-EUR-Lex. 2006. Available online: http://data.europa.eu/eli/reg/2006/1924/oj (accessed on 2 April 2024).
14. Rocha, A.F.; Silva, E.D.; de Araújo, M.G.G.; Germano, M.G.; Ferreira, P.C.; Bordin, V.; Ângelo, H.M. Preparation and Physical and Physical-Chemical Characterization of a Brownie enriched with Moringa Leaf Flour (Moringa Oleifera). *Res. Soc. Dev.* **2020**, *9*, e101973927. [CrossRef]

biology and life sciences forum

MDPI

Proceeding Paper

Changes in the Chemical Composition and Bioactive Compounds of Quinoa Seeds by Germination †

Jesica Romina Mufari [1,2], Patricia Paola Miranda-Villa [1,3], Antonella Estefania Bergesse [1,4], Andrea Carolina Rodríguez-Ruiz [1,5] and Edgardo Luis Calandri [1,3,*]

1 Instituto de Ciencia y Tecnología de los Alimentos (ICTA), Universidad Nacional de Córdoba (UNC), Av. Vélez Sarsfield 1611, Córdoba 5000, Argentina; rmufari@jla.com.ar (J.R.M.); patricia.miranda@mi.unc.edu.ar (P.P.M.-V.); abergesse@agro.unc.edu.ar (A.E.B.); acrodruiz@mi.unc.edu.ar (A.C.R.-R.)

2 Instituto de Investigaciones Biológicas y Tecnológicas (IIByT), CONICET, UNC, Av. Vélez Sarsfield 1611, Córdoba 5000, Argentina

3 Instituto de Ciencia y Tecnología de los Alimentos Córdoba (ICyTAC), CONICET, UNC, Av. Filloy S/N, Córdoba 5000, Argentina

4 Instituto Multidisciplinario de Biología Vegetal (IMBIV), CONICET, UNC, Av. Vélez Sarsfield 299, Córdoba 5000, Argentina

5 Instituto de Investigación y Desarrollo en Ingeniería de Procesos y Química Aplicada (IPQA), CONICET, UNC, Av. Vélez Sarsfield 1611, Córdoba 5000, Argentina

* Correspondence: edgardo.calandri@unc.edu.ar

† Presented at the VI International Congress la ValSe-Food, Lima, Peru, 23–25 September 2024.

Abstract: This research aimed to evaluate the changes that occur in the composition of macronutrients and soluble compounds of quinoa grains at different germination times. The seeds were soaked in water, drained, and then germinated in monolayers inside closed containers for 12, 24, 48, and 72 h and the germination was stopped by drying. Proteins, amino acids, fatty acids and antioxidant activity in flours were measured. A gradual reduction of carbohydrates is verified during the germination time with a concomitant increase in protein and lipid contents, while total minerals did not show modifications. The concentration effect due to metabolized carbohydrates seems responsible for the 33% rise in protein content 72 h after sprouting, but it is not enough to explain the almost 100% lipid increase for the same period. In general, amino acids and unsaturated fatty acids increase during germination, constituting a good resource for food and food ingredients intended for the general public, celiac patients, children, athletes, and elderly people.

Keywords: quinoa; germination; protein; amino acids; fatty acids; antioxidant activity

1. Introduction

Quinoa (*Chenopodium quinoa* Willd.) is a native plant of the Andean region of South America, which has similar chemical characteristics and uses to cereals but does not belong to the same botanical family, so it is commonly referred to as a pseudo-cereal. It was part of the basic food of the pre-Hispanic people of South America, though, after the Spanish conquest, the crop was marginalized in isolated areas and was mostly replaced by wheat and barley. It has an outstanding nutritional quality, mainly represented by the content and biological value of its proteins, the starch structure, the fat content (rich in unsaturated fatty acids), and the significant number of bioactive compounds such as fiber, minerals, vitamins, phytosterols, polyphenols, among others [1]. People with celiac disease can consume both the grain and its derived products. During germination an increase in soluble compounds such as amino acids, fatty acids, sugars, vitamins, and minerals takes place, favoring the bioavailability of these nutrients [2]. This process is stopped by heat or drying, ending the so-called malting. In recent years, several investigations have been carried out into gluten-free malted cereals and pseudo-cereals especially aimed at the celiac population.

In this work the changes in the macronutrients during the germination of quinoa grains are evaluated, improving the bioavailability of nutrients for celiac patients, but also for children, athletes, elderly people, and the general public.

2. Materials and Methods

2.1. Plant Material

The quinoa seeds (*Chenopodium quinoa* Willd.) were purchased from a quinoa producer from northwest Argentina. The seeds were cleaned and stored at 4 °C in sealed polyethylene bags until use.

2.2. Malted Quinoa Flours

Quinoa seeds were washed and moisturized in distilled water (1:10 ratio) for 2 h at 20 °C and then drained. Some of these seeds were dried in a fluidized bed dryer at 50 °C until moisture content was 11 ± 1% and then ground in a 0.25 mm hammer mill to obtain whole quinoa flour (QF0). The remaining seeds were arranged in a single layer on a plastic tray inside a closed container, away from direct sunlight. Germination was carried out at 25 °C and 50% relative humidity. The different stages of the process were monitored (Figure 1), stopping it at 12, 24, 48, and 72 h by drying and grinding following the same methodology described for QF0. Germination tests were carried out in triplicate, with the selection criteria being more than 90% sprouted grains [3].

12 h of germination | 24 h of germination

48 h of germination | 72 h of germination

Figure 1. Monitoring of the different stages of the germination process.

2.3. Physicochemical Characterization of the Flours

2.3.1. Proximate Analysis

Proximate analysis was performed in triplicate using AOAC techniques [4]. Carbohydrates (C) were calculated by difference. All determinations were expressed on a dry basis.

2.3.2. Total Amino Acids Profile

Flours were subjected to acid hydrolysis with 6M HCl under reflux for 24 h, following AOAC 994.12 methodology [4]. Amino acids identification and quantification were carried out using a high-performance liquid chromatography (HPLC) equipment with UV-detector, while data acquisition and processing was done using a Total Chrom Workstation software (version 6.3), following the methodology described in Mufari et al. [5].

2.3.3. In Vitro Protein Digestibility

In vitro protein digestibility (PD) of samples was determined according to the technique of Dierick et al. [6]. PD results were expressed as a percentage on a dry basis.

2.3.4. Total Fatty Acids Profile

Quinoa flours were soaked in n-hexane at a 1:10 ratio for 24 h at 3 °C in darkness to extract lipids. Following extraction, the mixture was filtered, the solvent was evaporated, and the fatty acids were trans-methylated. Identification and quantification of the fatty acids in the extracted oils were performed using gas chromatography with a mass spectrometry detector (GC-MS) on a Clarus 600 Perkin Elmer instrument, following the procedure outlined by Mufari et al. [5].

2.3.5. Evaluation of Antioxidant Properties

The extraction of antioxidant compounds was carried out according to [7]. Total phenolic content (TPC) was determined using the Folin-Ciocalteu method [8], adapted to microplate determinations. The absorbance of samples was measured in a spectrophotometer at 760 nm (BMG Labtech GmbH, Germany), and results were expressed as mg gallic acid equivalents (GAE)/100 g flour. Total flavonoid content (TFC) was determined according to the $AlCl_3$ method [8] adapted to a microplate reading absorbance at 367 nm, and the results were expressed in mg Quercetin equivalents (QE)/100 g flour. The antioxidant activity of the extracts was determined using 2,2-dyphenil-1-picryl hydrazyl (DPPH•) [8], absorbance was measured at 517 nm, and radical scavenging capacity (RSC) was calculated by means of the following equation: %RSC= [1 − (Absorbance (DPPH•) − Absorbance sample))/Absorbance (DPPH•)] ×100.

2.4. Statistical Analysis

Results were expressed as mean ± standard deviation of the replicates. Data analysis was carried out using InfoStat® professional (version 2020I). Significant differences between measurements were estimated by variance analysis (ANOVA). When statistically significant differences were observed (α = 0.05) a DGC multiple comparisons test was subsequently used.

3. Results and Discussion

Physico-Chemical Characterization of the Flours

Table 1 shows the results of the proximal analysis of the flours. Although moisture was adjusted to 11% prior to milling, the flours showed different degrees of hydration; hence, wet content was between 9.59 and 16.86%.

Table 1. Proximal composition and digestibility of the different quinoa flours.

Components	Germination Time				
	QF0	QF12	QF24	QF48	QF72
Moisture	9.59 ^{a} ± 0.23	16.49 ^{d} ± 0.01	13.24 ^{b} ± 0.01	14.33 ^{c} ± 0.01	16.86 ^{e} ± 0.14
Ash	1.97 ^{a} ± 0.02	1.68 ^{a} ± 0.10	2.29 ^{a} ± 0.13	2.05 ^{a} ± 0.11	1.89 ^{a} ± 0.26
Lipids	7.87 ^{a} ± 0.22	7.65 ^{a} ± 0.52	9.56 ^{b} ± 0.03	11.07 ^{c} ± 0.57	15.63 ^{d} ± 0.26
Proteins	13.56 ^{a} ± 0.30	17.19 ^{b} ± 0.10	17.38 ^{b} ± 0.25	18.34 ^{c} ± 0.10	18.03 ^{c} ± 0.16
Carbohydrates	76.60	73.48	70.77	68.54	64.45
% PD	71 ^{b} ± 2	72 ^{b} ± 1	81 ^{c} ± 1	82 ^{c} ± 1	62 ^{a} ± 2

The results are expressed in g/100 g of sample on a dry basis (except wet moisture content), means with standard deviations are reported (n = 3). Different letters in the same row denote statistically significant differences ($p < 0.05$). QF: quinoa flour, the number indicates the germination hours of the grains (0, 12, 24, 48, and 72 h); PD: protein digestibility.

The effect of sprouting on the nutritional composition of quinoa flour can be observed by the significant decrease of the total carbohydrate content (4 to 16% compared to QF0) with germination, a consequence of the metabolic activity of the seeds that takes place during germination. These results resemble the values reported for different varieties of sprouted rice, sorghum, millet, buckwheat, and amaranth, with decreases in carbohydrate content from 5 to 51% [9].

On the other hand, quinoa flours germinated for 48 and 72 h (QF48 and QF72) showed the highest protein values (Table 1). The synthesis of enzymes and non-protein nitrogenous substances (such as nucleic acids) can contribute to this increase. Diverse variations of protein contents are reported in sprouted cereals and pseudocereals, from reductions between 2 and 21% to increases between 5 and 100%. These differences depend on the plant material and on soaking, sprouting, and drying conditions [9]. In this investigation, an increase of 35% in protein content, between 0 and 72 h of germination, was observed, which implies an improvement in the nutritional profile. The same tendencies were found by Maldonado-Alvarado et al. [10] under similar conditions. A protein increase with carbohydrate reduction is good news for celiac patients commonly facing gluten-free foods that are highly energetic but poor in proteins.

Lipid contents were also increased with germination time, being almost double in QF72 compared to QF0 (Table 1).

In quinoa, lipids and protein are found mainly in the embryo while starch is in the perisperm. Amylases free glucose and maltose from perisperm starch, being the main energy source for germination; on the other hand, all the available nitrogen is endogenous as malting excludes any external source. So, the total nitrogen (as proteins) per 100 g existing at the beginning in QF0 should remain in the final stage 72 h later, then:

$$\frac{13.56}{(76.6-x)} = \frac{18.03}{64.45} \quad x = 28.2 \text{ g} \tag{1}$$

where x is the fraction of carbohydrates consumed by seed metabolism after 72 h of germination, nearly 37% of initial starch, showing the importance of carbohydrates for sprouting through providing the needed energy. The increase in lipids is greater than expected due to the concentration effect. Although some works show an opposite tendency for quinoa, Pachari Vera et al. [11] found the same behavior in four Peruvian quinoa varieties, in some cases with increases near 50%. The total mineral content did not show significant differences between germination stages (Table 1).

In vitro digestibility for proteins is presented in Table 1. The first 12 h of sprouting shows no significant changes, staying at 71–72%, but in the next 24–48 h it rises up to 81–82% and then drops to 62% after 72 h of germination (Table 1). A similar behavior was found by Prasad & Sahu [12] with values going from 72% to 87.5% after 48 h of sprouting, but these authors did not continue germination beyond.

Table 2 summarizes the results of the total amino acid composition of the different flours. As can be seen in Table 2, lysine contents in quinoa flour (QF0) are almost double those of wheat, corn, and rice while sulfur amino acids are in a similar proportion to those in the above cereals, but with quinoa protein content being higher its total contribution of amino acids is clearly superior. Some amino acids may diminish during germination as pointed out by Moongngarm and Saetung [13].

Table 2. Amino acid profile (g/100 g of flour) of integral quinoa flour (control) and germinated flours at different times.

Amino Acids	QF0	QF12	QF24	QF48	QF72
Aspartic ac.	0.442 [a] ± 0.028	0.973 [b] ± 0.094	0.954 [b] ± 0.076	1.067 [b] ± 0.008	1.090 [c] ± 0.013
Glutamic ac.	1.487 [a] ± 0.013	1.601 [c] ± 0.022	1.580 [b] ± 0.061	1.500 [a] ± 0.055	1.553 [b] ± 0.037
Serine	0.025 [a] ± 0.003	0.354 [b] ± 0.004	0.348 [b] ± 0.025	0.379 [c] ± 0.004	0.502 [d] ± 0.105
Histidine	0.033 [a] ± 0.004	0.761 [d] ± 0.022	0.596 [b] ± 0.000	0.660 [c] ± 0.003	0.769 [d] ± 0.171
Glycine	0.438 [a] ± 0.004	0.638 [c] ± 0.003	0.570 [b] ± 0.009	0.550 [b] ± 0.002	0.559 [b] ± 0.071
Threonine	0.071 [a] ± 0.003	0.508 [c] ± 0.018	0.430 [b] ± 0.010	0.485 [c] ± 0.008	0.560 [d] ± 0.009
Arginine	0.758 [a] ± 0.005	1.281 [d] ± 0.029	1.075 [c] ± 0.032	1.128 [c] ± 0.020	0.877 [b] ± 0.056
Alanine	0.654 [d] ± 0.001	0.550 [c] ± 0.010	0.512 [b] ± 0.003	0.498 [a] ± 0.003	0.554 [c] ± 0.017
Proline	0.048 [a] ± 0.015	0.188 [b] ± 0.046	nd	nd	nd
Tyrosine	0.130 [a] ± 0.015	0.601 [d] ± 0.014	0.463 [b] ± 0.042	0.513 [c] ± 0.099	0.580 [c] ± 0.076
Valine	0.571 [e] ± 0.020	0.144 [c] ± 0.005	0.107 [b] ± 0.008	0.086 [a] ± 0.061	0.553 [d] ± 0.004
Methionine ± Cysteine	5.360 [a] ± 0.019	5.994 [b] ± 0.106	6.936 [c] ± 0.171	6.816 [c] ± 0.227	6.091 [b] ± 1.026
Isoleucine	0.404 [a] ± 0.002	0.616 [d] ± 0.006	0.583 [c] ± 0.016	0.518 [b] ± 0.007	0.687 [d] ± 0.103
Leucine	0.561 [a] ± 0.002	0.872 [c] ± 0.016	0.796 [b] ± 0.023	0.776 [b] ± 0.002	1.056 [d] ± 0.146
Phenylalanine	0.253 [a] ± 0.002	0.800 [e] ± 0.006	0.640 [c] ± 0.011	0.573 [b] ± 0.034	0.733 [d] ± 0.110
Lysine	0.480 [a] ± 0.004	0.471 [a] ± 0.003	0.530 [b] ± 0.010	0.602 [c] ± 0.005	0.774 [d] ± 0.126
% of Recovery	92.44	95.10	92.72	86.74	93.82

QF: quinoa flour, the number indicates the germination hours of the grains (0, 12, 24, 48, and 72 h). Means with standard deviations are reported (SD, n = 6), expressed in g/100 g of sample on a dry basis. nd: below the limit of quantification (0.012). Different letters in the same row denote statistically significant differences ($p < 0.05$).

Table 3 shows the changes in the fatty acid profile of quinoa flour depending on the germination time, with palmitic (16:0), oleic (18:1ω9), and linoleic (18:2ω6) acids being the majority. Guardianelli et al. [14] found similar trends in germinated amaranth seeds, with increases in palmitic, linoleic, and linolenic acids and a decrease in stearic and oleic acids. The ω6/ω3 ratio, which is 9.6 for QF0, is reduced to 5.3 in QF72. This result is close to the recommended 5:1 [14].

Antioxidant compounds significantly increased with germination time, until 24–48 h of germination, and then at 72 h a decrease occurs, a similar trend to several of the parameters previously determined in this work. The TPC of quinoa flour was 54 ± 2 mg GAE 100 g^{-1} flour, a value that is within the range reported by other authors (25.0–71.7 mg GAE 100 g^{-1} flour) [7,15]. Total phenols increased by 63% after a 24-h germination.

On the other hand, the TFC of quinoa flour was higher than the reported by Carciochi & Dimitrov [7] (24 ± 1 mg QC 100 g^{-1} flour and 11 mg QC 100 g^{-1} flour, respectively). These differences can be partially explained by the differences in the extraction methods, the variety of the grains and the growing conditions. Total flavonoids increased by 125% after germination for 24 h, in comparison to raw quinoa. These increases in the compounds with antioxidant capacity, are related to enzymes activated during germination that lead to the release of compounds bound to cell structure of the matrix or by de novo synthesis [7]. Although other authors reported a constant increase with the germination time, in this study, antioxidant compounds decrease after 48 h of germination, probably due to the fixation in the new matrix of the growing seedling that makes their extraction difficult.

Antioxidant activity of germinated quinoa seeds extracts assessed by DPPH• method, indicated that the germination process significantly increased the antioxidant activity, compared to the control sample. Maximum radical scavenging activities were obtained in the lipid (89%) and ethanolic fraction (77%). Similar results were reported for different cereals and pseudo cereals, but with different optimal times of germination to obtain a maximum amount of soluble or bioavailable nutrients [9]. In the case of quinoa under the established malting conditions, germination time should be between 24 and 48 h.

Table 3. Fatty acid profile (g/100 g of flour) of integral quinoa flour (control) and germinated flours at different times.

Fatty Acids		QF0	QF12	QF24	QF48	QF72
Miristic	14:0	nd	$0.20^{a} \pm 0.01$	$0.23^{a} \pm 0.03$	$0.24^{a} \pm 0.01$	$0.22^{a} \pm 0.03$
Pentadecylic	15:0	nd	$0.14^{a} \pm 0.01$	$0.12^{a} \pm 0.02$	$0.14^{a} \pm 0.02$	$0.13^{a} \pm 0.01$
Palmitic	16:0	$9.2^{a} \pm 0.3$	$12.1^{c} \pm 0.2$	$12.5^{c} \pm 0.4$	$12.5^{c} \pm 0.2$	$11.3^{b} \pm 0.4$
Palmitoleic	16:1	nd	$0.41^{a} \pm 0.02$	$0.42^{a} \pm 0.03$	$0.38^{a} \pm 0.03$	$0.42^{a} \pm 0.02$
Margaric	17:0	nd	nd	$0.10^{a} \pm 0.03$	$0.11^{a} \pm 0.03$	$0.10^{a} \pm 0.03$
	17:1	nd	nd	$0.12^{a} \pm 0.03$	$0.12^{a} \pm 0.03$	$0.13^{a} \pm 0.03$
Stearic	18:0	$1.03^{c} \pm 0.03$	$0.90^{b} \pm 0.02$	$0.81^{a} \pm 0.02$	$0.79^{a} \pm 0.01$	$0.91^{b} \pm 0.03$
Oleic (ω9)	18:1	$27.6^{c} \pm 0.4$	$27.4^{c} \pm 0.5$	$24.6^{b} \pm 0.2$	$23.2^{a} \pm 0.5$	$26.8^{c} \pm 0.2$
Linoleic (ω6)	18:2	$54.9^{c} \pm 0.2$	$44.9^{a} \pm 0.4$	$44.3^{a} \pm 0.2$	$45.4^{b} \pm 0.3$	$45.8^{b} \pm 0.2$
Linolenic (ω3)	18:3	$5.7^{a} \pm 0.4$	$8.2^{b} \pm 0.4$	$9.8^{c} \pm 0.4$	$10.2^{c} \pm 0.4$	$8.9^{b} \pm 0.4$
Arachidonic	20:0	$0.33^{a} \pm 0.02$	$0.60^{b} \pm 0.03$	$0.78^{c} \pm 0.02$	$0.81^{c} \pm 0.02$	$0.58^{b} \pm 0.04$
Gondolic	20:1	$1.19^{a} \pm 0.01$	$2.49^{c} \pm 0.02$	$2.50^{c} \pm 0.02$	$1.72^{b} \pm 0.03$	$2.68^{d} \pm 0.03$
	20:2	nd	$0.18^{a} \pm 0.02$	$0.21^{a} \pm 0.01$	$0.20^{a} \pm 0.01$	$0.22^{a} \pm 0.03$
Behenic	22:0	nd	$0.98^{b} \pm 0.02$	$1.10^{b} \pm 0.03$	$1.11^{b} \pm 0.03$	$0.89^{a} \pm 0.02$
Erucic	22:1	nd	$1.91^{b} \pm 0.05$	$2.45^{c} \pm 0.02$	$2.71^{d} \pm 0.02$	$1.58^{a} \pm 0.03$
Tricosylic	23:0	nd	$0.10^{a} \pm 0.02$	$0.1^{a} \pm 0.01$	$0.11^{a} \pm 0.02$	nd
Lignoceric	24:0	nd	$0.42^{a} \pm 0.02$	$0.58^{b} \pm 0.02$	$0.60^{b} \pm 0.03$	$0.41^{a} \pm 0.01$
	24:1	nd	$0.18^{a} \pm 0.04$	$0.20^{a} \pm 0.03$	$0.21^{a} \pm 0.04$	$0.19^{a} \pm 0.02$
Saturated (S)		10.56	15.44	16.33	16.41	14.54
Monounsaturated (MI)		28.79	32.39	30.30	28.34	31.80
Poliunsaturated (PI)		60.6	53.28	54.31	55.80	54.92
ω6/ω3		9.63	5.47	4.52	4.45	5.15
PU/MU		2.10	1.64	1.79	1.97	1.73

QF: quinoa flour, the number indicates the germination hours of the grains (0, 12, 24, 48, and 72 h). Means with standard deviations are reported (SD, n = 6), expressed in g/100 g of sample on a dry basis. nd = below the limit of quantification (0.012). Different letters in the same row denote statistically significant differences ($p < 0.05$).

4. Conclusions

It has been shown that it is feasible to increase the nutritional quality of quinoa grains by germinating them for a period between 24 and 48 h. Flours obtained from sprouted grains demonstrated a greater bioavailability of nutrients. The preliminary results obtained in this work confirm that germination is an adequate process for improving the nutritional profile of quinoa flours, and for further production of healthy foods intended for vulnerable age groups such as children or the elderly.

Author Contributions: J.R.M.: conceptualization, formal analysis, methodology, writing—original draft. P.P.M.-V.: designed the experiment, data curation, interpreted the results and wrote the manuscript. A.C.R.-R.: formal analysis, methodology, writing—original draft. A.E.B. Bergesse: Data curation, interpreted the results and wrote the manuscript. E.L.C. Calandri: project administration, funding acquisition, review and editing. All authors have read and agreed to the published version of the manuscript.

Funding: This work was supported by a grant from Ia ValSe Food-CYTED (Ref. 119RT0567) Universidad de Lima, the National Council for Scientific and Technical Research (CONICET) and the Secretariat of Science and Technology (SECyT) (Ref. 30720130101068CB).

Data Availability Statement: Data is contained within the article.

Acknowledgments: The authors express their gratitude to Universidad de Lima, the National Council for Scientific and Technical Research (CONICET) and the Secretariat of Science and Technology (SECyT) for funding as well as to the National University of Córdoba (UNC), ICTA, and ICyTAC for lending working spaces and offering available equipment used in this work.

Conflicts of Interest: The authors declare that they have no known competing financial interests or personal relationships that could have appeared to influence the work reported in this paper.

References

1. Vega-Gálvez, A.; Miranda, M.; Vergara, J.; Uribe, E.; Puente, L.; Martínez, E.A. Nutrition facts and functional potential of quinoa (*Chenopodium quinoa* willd.), an ancient Andean grain: A review. *J. Sci. Food Agric.* **2010**, *90*, 2541–2547. [CrossRef] [PubMed]
2. Edney, M.J.; Izydorczyk, M.S. Malt types and products. In *Encyclopedia of Food Sciences and Nutrition*, 3rd ed.; Caballero, B., Ed.; Academic Press: Cambridge, MA, USA, 2003; pp. 3671–3677.
3. Miranda-Villa, P.P.; Mufari, J.R.; Bergesse, A.E.; Calandri, E.L. Effects of Whole and Malted Quinoa Flour Addition on Gluten-Free Muffins Quality. *J. Food Sci.* **2019**, *84*, 147–153. [CrossRef] [PubMed]
4. Association of Official Analytical Chemists (AOAC). *Official Methods of Analysis of AOAC International*, 17th ed.; AOAC International: Gaithersburg, MD, USA, 2000.
5. Mufari, J.R.; Miranda-Villa, P.P.; Calandri, E.L. Quinoa germ and starch separation by wet milling, performance and characterization of the fractions. *LWT* **2018**, *96*, 527–534. [CrossRef]
6. Dierick, N.; Vervaeke, I.; Decuypere, J.; Hendrickx, H. *Protein Digestion in Pigs Measured In Vivo and In Vitro*; Just, A., Jørgensen, H., Fernández, J.A., Eds.; Digestive Physiology in the Pig; National Institute of Animal Science: Copenhagen, Denmark, 1985; pp. 329–332.
7. Carciochi, R.A.; Dimitrov, K.; Galván D'Alessandro, L. Effect of malting conditions on phenolic content, Maillard reaction products formation, and antioxidant activity of quinoa seeds. *J. Food Sci. Technol.* **2016**, *53*, 3978–3985. [CrossRef] [PubMed]
8. Bergesse, A.E.; Asensio, C.M.; Quiroga, P.R.; Ryan, L.C.; Grosso, N.R.; Nepote, V. Microencapsulation of phenolic compounds extracted from soybean seed coats by spray-drying. *J. Food Sci.* **2023**, *88*, 4457–4471. [CrossRef] [PubMed]
9. Omary, M.B.; Fong, C.; Rothschild, J.; Finney, P. Effects of Germination on the Nutritional Profile of Gluten-Free Cereals and Pseudocereals: A Review. *Cereal Chem.* **2012**, *89*, 1–14. [CrossRef]
10. Maldonado-Alvarado, P.; Pavón-Vargas, D.J.; Abarca-Robles, J.; Valencia-Chamorro, S.; Haros, C.M. Effect of germination on the nutritional properties, phytic acid content, and phytase activity of quinoa (*Chenopodium quinoa* Willd). *Foods* **2023**, *12*, 389. [CrossRef] [PubMed]
11. Pachari Vera, E.; Alca, J.J.; Rondón Saravia, G.; Callejas Campioni, N.; Jachmanián Alpuy, I. Comparison of the lipid profile and tocopherol content of four Peruvian quinoa (*Chenopodium quinoa* Willd.) cultivars ('Amarilla de Marangani', 'Blanca de Juli', INIA 415 'Roja Pasankalla', INIA 420 'Negra Collana') during germination. *J. Cereal Sci.* **2019**, *88*, 132–137. [CrossRef]
12. Prasad, P.; Sahu, J.K. Effect of soaking and germination on grain matrix and glycaemic potential: A comparative study on white quinoa, proso and foxtail millet flours. *Food Biosci.* **2023**, *56*, 103105. [CrossRef]
13. Moongngarm, A.; Saetung, N. Comparison of chemical compositions and bioactive compounds of germinated rough rice and brown rice. *Food Chem.* **2010**, *122*, 782–788. [CrossRef]
14. Guardianelli, L.M.; Salinas, M.V.; Puppo, M.C. Chemical and thermal properties of flours from germinated amaranth seeds. *J. Food Meas. Charact.* **2019**, *13*, 1078–1088. [CrossRef]
15. Alvarez-Jubete, L.; Wijngaard, H.; Arendt, E.K.; Gallagher, E. Polyphenol composition and in vitro antioxidant activity of amaranth, quinoa buckwheat and wheat as affected by sprouting and baking. *Food Chem.* **2010**, *119*, 770–778. [CrossRef]

biology and life sciences forum

MDPI

Proceeding Paper

Evaluation of a Functional Bread Made with Wheat Flour (*Triticum* spp.), Tarwi Flour (*Lupinus mutabilis* Sweet) and Hydroxypropyl Methyl Cellulose [†]

Paola Sánchez-Guamba and Pedro Maldonado-Alvarado *

Department of Food Science and Biotechnology, Escuela Politécnica Nacional, Quito 17-01-2759, Ecuador; paola.sanchez02@epn.edu.ec

* Correspondence: pedro.maldonado@epn.edu.ec

† Presented at the VI International Congress la ValSe-Food, Lima, Peru, 23–25 September 2024.

Abstract: Making bread from tarwi represents a scientific and technical challenge due to its poor breadmaking properties; however, strategies like partial substitution with other flours as well as the use of improvers could maximize these properties. This work aimed to evaluate a functional bread made with wheat flour (*Triticum* spp.), tarwi flour (*Lupinus mutabilis* Sweet) and Hydroxypropyl Methyl Cellulose (HPMC). The partial substitution levels evaluated were 0, 5, 10, 15, 20 and 25% with and without the addition of HPMC as a bread improver. The bread's volume, weight, height and specific volume were determined. Sensory tests of the bread were carried out to determine bakery aptitude and acceptability. The specific volumes of the bread were 3.90; 3.46; 3.27; 3.23; 2.98 and 2.62 cm^3/g in samples without HPMC and 4.60; 4.28; 4.27; 3.26; 2.96 and 2.62 cm^3/g in samples with HPMC, respectively. These values had reciprocity with the heights of the breads, which were between 8.90 and 6.00 cm in samples without HPMC and between 10.55 and 6.50 cm in samples with HPMC. The specific volume of bread decreased from 15% substitution. The sample with 0, 5 and 10% substitution with tarwi flour had both acceptability and bakery aptitude

Keywords: bread; tarwi; wheat; HPMC; sensory properties

1. Introduction

The marginalization of Andean crops has occurred due to different problems:=, including a low social prestige of endemic crops, a preference for the use of other imported crops (e.g., wheat), and a lack of public food sovereignty policies in southern countries, among others [1]. In addition, the valorization of Andean crops has not been easy because of the overproduction of crops, which can be understood as the non-use of a certain part of the food production, as well as poor valorization of endemic foods, related to the low valorization of raw materials in finished products [2]. However, it is well known that Andean crops, e.g., quinoa, amaranth, cassava, banana and lupin, have higher nutritional and functional properties, and can provide important benefits to human health [1].

In the Andes region, few studies related to research and development of new foods from the local matrix with functional properties such as tarwi, but also with maximized structural and sensory characteristics, have been reported. This is due to the difficulty of combining all these properties in synergy to obtain a finished product and, crucially, a gluten-free food [3]. The expanding number of gluten-free goods, like bread and extruded foods, use food matrices that are different from those that contain gluten, using products like pseudocereals, tubers and pulses to mimic the viscoelastic characteristics of this protein. Improvers like hydrocolloids, enzymes and modified starch can be added to achieve the desired expanded gluten-free network, which represents a technological challenge. On the other hand, achieving improved sensory characteristics with a gluten-free matrix is even more complex. Many of the gluten-free products that are on the market have a poor sensory perception and, on the other hand, people are accustomed to the taste of products that contain wheat, such as bread [4].

The aim of this contribution was to evaluate a functional bread made with wheat flour (*Triticum* spp.), tarwi flour (*Lupinus mutabilis* Sweet) and Hydroxypropyl Methyl Cellulose (HPMC).

2. Materials and Method

2.1. Materials

Dry seeds of tarwi (*Lupinus mutabilis* Sweet) from Latacunga, Ecuador, were used. The seeds were soaked in water for a period of 1 day, cooked at a temperature of 87 °C for 1 h and debittered in water for 7 days, with 3 water changes being carried out daily. A manual selection was carried out and the non-hydrated grains were discarded; the selected grains were dried in a Selecta model oven (Barcelona, Spain) at 45 °C for 48 h. The grinding was carried out in an Alpine model mill (Augsburg, Germany) for a time of 10 s and was sieved through a No. 11 mesh. YA brand (Quito, Ecuador) fortified wheat flour for baking was acquired from a local market. The additive used to improve the bread was Methocel E5LV, Hydroxypropylmethylcellulose (The Dow Chemical Company, Midland, MI, USA).

To make the bread, fortified wheat flour for baking, fresh yeast, multipurpose vegetable margarine, white sugar and iodized salt were used.

2.2. Breadmaking Test

Following the INEN 530-2013 Standard [5] (Ecuador), a baking test was carried out using the two-fermentation method and bread samples with different substitutions (substitutions of lupine flour with 0, 5, 10, 15, 20 and 25% of wheat flour were performed) with the addition or not of HPMC 2% (based on flour content). The amount of water that was added to the flour to form a dough of suitable consistency was based on the percentage of water absorption obtained from the Mixolab analysis and was replicated for all samples. The salt, sugar and yeast were dissolved in water at 30 ± 2 °C. The flour, HPMC (if established in the recipe) and margarine were placed in a deep container, and the water, salt and yeast solution were added. They were homogenized in a KENWOOD mixer (Tokyo, Japan), at a low speed (60 rpm) for 3 min, the dough was placed on baking trays and left to rest in a UNOX brand fermentation chamber, model XLT 135 (Rome, Italy), for 15 min. at 30 °C and a relative humidity of 80%. The dough was then kneaded and divided into 135 g portions, before being laminated, rolled and placed in the previously greased molds. Fermentation was carried out for a second time with the same characteristics but for a period of 45 min; the samples were cooked in a UNOX model XFT 115 forced convection oven (Rome, Italy) with the following conditions: 180 °C for 5 min with 80% humidity; 220 °C for 5 min without humidity; and 150 °C for 20 min without humidity and cooled until reaching an internal temperature of 30 °C. The weights, heights and volumes of the obtained breads were determined with a scale, a ruler and a pycnometer using birdseed. Specific volume was calculated as the ratio between the bread volume and the bread weight (mL/g).

2.3. Sensory Analysis

2.3.1. Baking Aptitude

A panel of 20 semi-trained judges was established, who evaluated the baking aptitude of the samples with the following parameters: crust color (15), appearance and symmetry of the bread (15), flavor (10), crumb color (10), crumb texture (30) and crumb grain (20) (maximum scores are indicated in parentheses). For analysis, 1 cm wide slices of bread from each substitution level were selected and placed in white containers with a 3-digit random code for each sample. The INEN standard 530-2013 was taken as a reference.

2.3.2. Acceptability

To understand the panel's preference for bread samples with different levels of substitution, 20 semi-trained judges evaluated the bread samples with a hedonic scale with values between 0 and 5, in which 0 meant "Very Bad", 1 "Bad", 2 "Fair", 3 "Acceptable", 4 "Good" and 5 "Very Good". The parameters evaluated were color, aroma, crumb texture, crust texture, flavor and symmetry. The sample preparation was carried out in the same way as in the baking aptitude evaluation [5].

2.4. Statistical Analysis

Results were expressed as mean ± standard deviation, and all measurements were conducted in triplicate. The results obtained were evaluated with an analysis of variance (ANOVA) followed by an LSD test with a confidence level of 95% in the STATGRAPHICS ® CENTURION XVI software.

3. Results and Discussion

3.1. Breadmaking Properties

Table 1 shows the results obtained for the different samples made with 0, 5, 10, 15, 20 and 25% of partial substitution of wheat flour with tarwi flour. In treatments without adding HPMC, no significant differences were found between the specific volume values of the 5, 10 and 15% replacement samples. For the different levels, the specific volume decreased by 11.3; 16.2; 17.2; 23.6 and 32.8%, respectively, compared to the value of bread with 0% substitution. On the other hand, a decrease in bread height was observed as the level of substitution increased because the CO_2 produced in fermentation is not retained, although there were significantly no differences between the different levels.

Table 1. Physical characteristics of bread with 0, 5, 10, 15, 20 and 25% partial replacement of wheat flour with tarwi flour.

Partial Substitution of Wheat Flour by Tarwi Flour	Weigth (g)	Volume (cm^3)	Specific Volume (cm^3/g)	Heigth (cm)
0%	$109.87^{a} \pm 1.56$	$429^{f} \pm 2.98$	$3.90^{d} \pm 0.05$	$8.90^{b} \pm 0.05$
0% *	$105.43^{a} \pm 1.56$	$485^{e} \pm 2.98$	$4.60^{e} \pm 0.05$	$10.55^{bc} \pm 0.05$
5%	$110.15^{a} \pm 1.02$	$381^{d} \pm 3.98$	$3.46^{c} \pm 0.23$	$7.90^{a} \pm 0.84$
5% *	$109.80^{b} \pm 1.02$	$469^{d} \pm 3.98$	$4.28^{d} \pm 0.23$	$10.00^{bc} \pm 0.84$
10%	$116.08^{c} \pm 1.83$	$379.5^{d} \pm 3.87$	$3.27^{c} \pm 0.03$	$7.75^{a} \pm 1.03$
10% *	$109.33^{b} \pm 1.83$	$467^{d} \pm 3.87$	$4.27^{d} \pm 0.03$	$9.90^{b} \pm 1.03$
15%	$114.12^{b} \pm 1.63$	$369^{c} \pm 1.35$	$3.23^{c} \pm 0.07$	$7.35^{a} \pm 1.47$
15% *	$113.47^{c} \pm 1.63$	$371^{c} \pm 1.35$	$3.26^{c} \pm 0.07$	$8.10^{ab} \pm 1.47$
20%	$113.62^{b} \pm 1.17$	$339^{b} \pm 5.72$	$2.98^{b} \pm 0.09$	$7.05^{a} \pm 2.09$
20% *	$115.37^{c} \pm 1.17$	$312^{b} \pm 5.72$	$2.96^{b} \pm 0.09$	$7.30^{a} \pm 2.09$
25%	$112.42^{b} \pm 2.31$	$294^{a} \pm 3.67$	$2.62^{a} \pm 0.03$	$6.00^{a} \pm 2.73$
25% *	$115.52^{c} \pm 2.31$	$299^{a} \pm 3.67$	$2.62^{a} \pm 0.03$	$6.50^{a} \pm 2.73$

$\overline{x} \pm \sigma$ (n = 4). Different letters within the column indicate significant differences at $p < 0.05$ (LSD). * Samples with the addition of 2% HPMC based on flour content.

In treatments with the addition of HPMC, a significant increase in the specific volume was observed in the samples with 0, 5 and 10% replacement with lupine flour. The percentage increases were, respectively, 17.95; 23.21 and 30.58%. However, for the 15, 20 and 25% samples, no changes were observed in the specific volume, and the samples' original values were maintained.

Alasino et al. (2011) [6] showed that the average specific volume of bread made from wheat flour ranges around 3.98 cm^3/g and that lower values occur in flours considered weak. According to Mongi et al. [7] in their study of making breads with soy flour, as the substitution percentage increases, the specific volume of the bread decreases. With a substitution level of 20%, a specific volume of 3 cm^3/g is obtained, this is because there is no adequate protein source, and it does not allow for the appropriate retention of CO_2 to increase the volume of the bread. Rosell et al. [8] reported a specific volume of about 3 cm^3/g for partial substitutions of wheat flour with 0 and 12.5% of tarwi flour, but the specific volume of bread decreased to 2, 1.2 and 1 cm^3/g for 25, 50 and 100% of substitution with tarwi.

3.2. Sensory Analysis

Table 2 shows the results obtained for the different samples made with 0, 5, 10, 15, 20 and 25% of partial substitution of wheat flour with tarwi flour. In treatments without the addition pf HPMC, it was determined that samples with up to 10% substitution obtained scores greater than 80. Samples with a higher percentage of substitution obtained scores less than 70. The evaluated parameters did not present statistically significant differences between the samples with 5, 10, 15, 20 and 25% of substitution. The results of the evaluated parameters showed a relationship directly proportional to the percentage of bread substitution. However, in the flavor parameter, the values decreased drastically, due mainly to the fact that the bread took on a strange flavor as the substitution percentage increased.

Table 2. Baking ability for breads made with 0, 5, 10, 15, 20 and 25% partial replacement of wheat flour with lupine flour.

Partial Substitution of Wheat Flour by Tarwi Flour	Sensory Parameters for Baking Aptitude						
	Crust Color (/15)	Appearance and Symmetry (/15)	Flavor (/10)	Crumb Color (/10)	Crumb Texture (/30)	Crumb Grain (/20)	Sum (/100)
0%	$14.50^{a} \pm 1.41$	$14.20^{b} \pm 1.45$	$8.70^{c} \pm 1.55$	$9.40^{b} \pm 1.78$	$27.50^{b} \pm 3.69$	$18.40^{a} \pm 2.73$	$92.70^{b} \pm 8.65$
0% *	$14.20^{a} \pm 1.41$	$14.50^{b} \pm 1.45$	$8.30^{b} \pm 1.55$	$9.50^{b} \pm 1.78$	$28.50^{b} \pm 3.69$	$18.30^{a} \pm 2.73$	$93.30^{b} \pm 8.65$
5%	$14.30^{a} \pm 1.58$	$14.00^{ab} \pm 2.53$	$8.10^{bc} \pm 2.40$	$9.10^{b} \pm 2.11$	$26.30^{b} \pm 2.72$	$17.20^{a} \pm 2.11$	$89.00^{ab} \pm 9.12$
5% *	$14.10^{a} \pm 1.58$	$14.10^{b} \pm 2.53$	$8.10^{b} \pm 2.40$	$9.20^{b} \pm 2.11$	$27.30^{b} \pm 2.72$	$17.10^{a} \pm 2.11$	$89.90^{ab} \pm 9.12$
10%	$13.90^{a} \pm 3.79$	$13.80^{ab} \pm 2.78$	$7.90^{ab} \pm 2.73$	$7.60^{ab} \pm 2.10$	$25.50^{b} \pm 2.96$	$16.40^{a} \pm 1.26$	$85.10^{ab} \pm 11.71$
10% *	$13.80^{a} \pm 3.79$	$13.70^{b} \pm 2.78$	$7.90^{b} \pm 2.73$	$8.60^{ab} \pm 2.10$	$26.50^{b} \pm 2.96$	$16.20^{a} \pm 1.26$	$86.70^{ab} \pm 11.71$
15%	$10.70^{a} \pm 4.56$	$10.90^{ab} \pm 3.38$	$3.50^{ab} \pm 2.85$	$7.20^{ab} \pm 1.45$	$20.40^{ab} \pm 3.69$	$16.10^{a} \pm 3.41$	$68.80^{a} \pm 13.33$
15% *	$10.50^{a} \pm 4.56$	$10.40^{ab} \pm 3.38$	$3.30^{a} \pm 2.85$	$7.10^{a} \pm 1.45$	$19.40^{a} \pm 3.69$	$15.90^{a} \pm 3.41$	$66.60^{a} \pm 13.33$
20%	$10.30^{a} \pm 6.34$	$9.70^{a} \pm 3.53$	$3.10^{a} \pm 2.03$	$6.30^{a} \pm 0.98$	$20.10^{a} \pm 6.46$	$15.90^{a} \pm 5.76$	$65.40^{a} \pm 15.15$
20% *	$10.10^{a} \pm 6.34$	$9.60^{a} \pm 3.53$	$3.10^{a} \pm 2.03$	$6.10^{a} \pm 0.98$	$19.10^{a} \pm 6.46$	$15.70^{a} \pm 5.76$	$63.70^{a} \pm 15.15$
25%	$9.80^{a} \pm 6.45$	$8.60^{a} \pm 3.62$	$3.00^{a} \pm 3.11$	$6.00^{a} \pm 2.06$	$18.90^{a} \pm 4.08$	$14.80^{a} \pm 6.36$	$61.10^{a} \pm 17.16$
25% *	$9.50^{a} \pm 6.45$	$8.50^{a} \pm 3.62$	$3.00^{a} \pm 3.11$	$6.00^{a} \pm 2.06$	$18.90^{a} \pm 4.08$	$14.50^{a} \pm 6.36$	$60.40^{a} \pm 17.16$

$\bar{x} \pm \sigma$ (n = 4). Different letters within the column indicate significant differences at $p < 0.05$ (LSD). * Samples with the addition of 2% HPMC based on flour content.

According to Zuleta et al. [9], breads with a partial substitution of soy flour obtained similar results; it was observed that the breads with 0, 5 and 10% of substitution had the best baking aptitude.

In treatments with the addition of HPMC, it was determined that samples with up to 10% substitution obtained scores greater than 80. The evaluated parameters did not present a statistically significant difference between the samples with 5, 10, 15 and 20% of substitution. The results of the evaluated parameters indicated that the breads with the best characteristics were those with 0, 5 and 10% substitution with lupine flour. The addition of HPMC improved the evaluated parameters since it provided the breads with superior sensory characteristics [10].

4. Conclusions

Andean farinaceous like tarwi can be used to obtain new functional foods, e.g., bread with maximized breadmaking and sensory properties. The addition of HPMC 2% to the doughs with partial substitution of wheat with tarwi can increase the specific volume of the bread for up to 15% of substitution.

Author Contributions: Conceptualization, P.M.-A.; methodology, P.M.-A. and P.S.-G.; validation, P.M.-A.; formal analysis, P.S.-G.; investigation, P.S.-G.; writing—original draft preparation, P.M.-A.; writing—review and editing, P.M.-A.; funding acquisition, P.M.-A. All authors have read and agreed to the published version of the manuscript.

Funding: This research was funded by PIS-21-04 project of EPN from Quito-Ecuador.

Institutional Review Board Statement: Not applicable.

Informed Consent Statement: Not applicable.

Data Availability Statement: The data analyzed in this study are available from the authors upon reasonable request.

Acknowledgments: This work was supported by grant Ia ValSe Food-CYTED (Ref. 119RT0567), Universidad de Lima—Peru and EPN from Quito-Ecuador for financing support.

Conflicts of Interest: The authors declare no conflict of interest.

References

1. Haros, C.M.; Reguera, M.; Sammán, N.; Paredes-López, O. *Latin-American Seeds: Agronomic, Processing and Health Aspects*; CRC Press: Boca Raton, FL, USA, 2023; pp. 1–442. [CrossRef]
2. Walsh-Dilley, M. Resilience Compromised: Producing Vulnerability to Climate and Market among Quinoa Producers in Southwestern Bolivia. *Glob. Environ. Chang.* **2020**, *65*, 102165. [CrossRef]
3. Gallagher, E. *Gluten-Free Food Science and Technology*; Wiley: Hoboken, NJ, USA, 2009; pp. 1–246. [CrossRef]
4. Capriles, V.; Areas, A. Novel Approaches in Gluten-Free Breadmaking: Interface between Food Science, Nutrition, and Health. *Compr. Rev. Food Sci. Food Saf.* **2014**, *13*, 871–890. [CrossRef]
5. *NTE INEN 530:2013*; I.N.E.N. Instituto Ecuatoriano de Normalización. Harina de Trigo. Ensayo de Panificación: Quito, Ecuador, 2013.
6. Alasino, N.M.; Aringoli, E.E.; Sánchez, H.D. Optimización de Los Niveles de Los Agentes Oxidante/Reductor En La Panificación de Harina de Trigo. *Inf. Tecnológica* **2011**, *22*, 75–86. [CrossRef]
7. Mongi, R.; Ndabikunze, B.; Chove, B.; Mamiro, P.; Ruhembe, C.C.; Ntwenya, J. Proximate Composition, Bread Characteristics and Sensory Evaluation of Cocoyam-Wheat Composite Breads. *Afr. J. Food, Agric. Nutr. Dev.* **2011**, *11*, 5586–5599. [CrossRef]
8. Rosell, C.M.; Cortez, G.; Repo-Carrasco, R. Breadmaking Use of Andean Crops Quinoa, Kañiwa, Kiwicha, and Tarwi. *Cereal Chem. J.* **2009**, *86*, 386–392. [CrossRef]
9. Zuleta, A.; Binaghi, M.; Greco, C.; Aguirre, C.; De la Casa, L.; Tadini, C.; Ronayne, P. Diseño de Panes Funcionales a Base de Harinas No Tradicionales. *Rev. Chile Nutr.* **2012**, *39*, 58–64. [CrossRef]
10. San Miguel Fermoso, M.Á. Influencia de La Hidroxipropilmetilcelulosa, El Psyllium y Su Combinación En Las Propiedades Morfogeométricas y Texturales de Panes Sin Gluten Elaborados Con Harina de Arroz. Master's Thesis, Universidad de Valladolid, Valladolid, Spain, 2013.

biology and life sciences forum

MDPI

Proceeding Paper

Curcubita moschata Seeds: Ancestral Flavor and Nutrition for Current Use

Elsa Julieta Salazar * and María Eunice Cota

Escuela de Nutrición, Facultad de Ciencias Químicas y Farmacia and Universidad de San Carlos de Guatemala, Guatemala City 01012, Guatemala; eunisnutr49@gmail.com
* Correspondence: salazarjulieta2021@profesor.usac.edu.gt

Abstract: The squash, *C. moschata*, is a type of pumpkin that grows easily in milpas—small, polyculture fields in Guatemala. Excavations carried out in pre-Columbian mounds in Uaxactún, Petén, indicate that squash has been cultivated and consumed for more than 5000 years. Today, both the pulp and the seed are still used as food; seeds are processed by hand and sold as toasted seed, or as toasted and ground seed with added salt, which is called *pepita*. The seed is used as a flavoring and thickening ingredient in sauces and pepita is used as an accompaniment to fresh fruit. This work aimed to provide updated information on the nutritional compositions of squash seeds and *pepita* in popular markets in the north, center, west, and southeast of Guatemala. The moisture content was determined in a convection oven at 60 °C, the ash by combustion in furnace at 450 °C, the protein by Kjeldahl method, fat by solvent extraction in the goldfish apparatus, and minerals by atomic absorption spectroscopy; and a UV/VIS colorimeter. The moisture content in seeds and *pepita* were 5.67% and 4.65%; ash 4% and 6.24%; protein 32.9% and 29.21%; lipids 32.07% and 30.22%, respectively. There was a higher content of macrominerals in *pepita* than in seeds, due to the salt addition. Comparing the nutritional results of the dry seed with those reported in the Food Composition Table for Central America, differences in protein and fat content are greater than 10%.

Keywords: *Curcubita moschata*; macronutrients; minerals; nutrition; seeds

1. Introduction

Ayote is a type of squash that grows as a creeping plant in corn fields in Guatemala. It belongs to the Cucurbitaceae family, and its scientific name is *Cucurbita moschata* Duchesne. Guatemala is the point of origin of *Cucurbita moschata* D [1]; there are archaeological records that mention *C. moschata* as a plant cultivated and consumed in Guatemala for more than 5000 years [2]; in excavations carried out in pre-Columbian mounds, in Uaxactún, Petén, researchers found carbonized peduncles of *C. moschata* [2].

The cultivation of squash is part of the Mesoamerican food system known as milpa, which includes the cultivation of corn, beans, and squash in the same physical space and in biological harmony, which allows the spontaneous growth of a variety of herbs with edible leaves and stems that, together, are called *quiletes* [3].

Both the flowers and the tender stems of the ayote, the immature fruits, the mature fruits, and its seeds are edible. Squash seeds are abundant, extracted from the ripe fruit and easily dried in the sun or at room temperature. The seeds are usually toasted and consumed as a snack, but they can also be ground and added as a thickening agent to sauces. The sauce known as "*iguashte*" in Spanish, or Ch'ereb'an in Kaqchiquel, is prepared from these seeds; it is a thick sauce obtained from squash seeds blended in the water used to boil meats or vegetables, seasoned with salt and chili. Approximately one ounce of toasted squash seeds is used per serving of vegetables or meat [3].

Another popular way in which squash seeds are consumed is as "*pepita*". *Pepita* is prepared using artisanal methods: people toast them on a clay or metal pan, grind them in a disc mill, and then add salt. The most frequent use of the *pepita* is to accompany pieces of fresh fruit, which are sold on the street. Normally, a teaspoon of *pepita* is added per 250 g portion of fruit.

Due to the importance of seeds in the diet, the frequent use of squash seeds in the Guatemalan diet and the lack of information on *pepita's* nutritional composition, we present this study to provide more data on the nutritional composition of the squash seed and the *pepita* sold in popular markets in five regions of Guatemala.

2. Materials and Methods

2.1. Sample Collection

The squash seeds and *pepita* were obtained in bulk from five markets of Guatemala: Flores and Santa Cruz del Quiché (northern region), Jutiapa (southern region), Jacaltenango (western region), and Santiago Atitlán, (central region).

2.2. Macronutrient and Minerals Quantification

The proximal composition was determined by official techniques. The analysis matrix was dried and pulverized squash seed, and dried and pulverized *pepita*, both with a particle size of 0.5 mm. The humidity was determined in a convection oven at 60 °C until it was a constant weight; the ash by muffle at 450 °C, the raw protein by the Kjeldhal method using Kjeltec Auto 1030 Analyzer equipment; the lipids by petroleum benzene extraction using Goldfish equipment. The minerals were determined by atomic absorption spectroscopy, using Perkin Elmer AAnalyst 100 equipment, except phosphorus, which was determined using a UV/VIS Lambda 11 colorimeter. The analyses were conducted in triplicate.

2.3. Statistical Analysis

Macronutrient and mineral content results were expressed as mean +/− standard deviation. The results were analyzed using descriptive statistics and analysis of variance -ANOVA + post hoc to determine statistically significant differences, through the Statgraphics Centurion19.

3. Results and Discussion

3.1. Macronutrients and Minerals in Cucurbita Moschata Seeds

Table 1 shows the macronutrients and mineral content of *C. moschata* seeds from different regions. Squash seeds had high protein and lipid and low water content. This is characteristic of all seeds, since they must accumulate reserve nutrients for germination purposes [4].

The protein content was in the range of 30 to 35%. Squash seeds from Flores, Petén had the lowest value. This amount of protein is similar to that reported in code 10015 of the Central American Food Composition Table, and in code 12014 of the USDA database, for dried squash seeds and raw pumpkin seeds, respectively [5,6].

The average lipid content was 31.44% and the range was between 29.31% and 32.78%. The lowest content was found in squash seeds from Jacaltenango. In general, the amount of lipids in squash seeds was high, but it was observed that there was 10 to 15% lower lipid content compared to data from the food composition tables and databases [5,6]. Alfawaz [7] and Manda Devi, Prassad and Palmei [8] analyzed the nutritional composition of whole pumpkin seeds and kernels and found 28–32% of lipids in whole seeds and 32–44% in kernels. Petkova and Antova [9] studied the protein and lipid content in *C. moschata* seeds at 30, 60, and 90 days of post-flowering growth, and found that the concentration of both nutrients is directly related to the post-flowering time. This could explain the difference found in this study, given that the seeds analyzed were whole seeds, and probably the maturity level of the fruit from which the seeds were extracted was around 45 days post-flowering growth.

Table 1. Macronutrients (g/100 g) and minerals (mg/100 g) in *C. moschata* seeds.

	Jutiapa	Flores, Petén	Santa Cruz, Quiché	Jacaltenango	Santiago Atitlán	Media
Water	3.61 ± 0.09 [a]	4.79 ± 0.03 [b]	5.34 ± 0.19 [c]	6.47 ± 0.09 [d]	6.56 ± 0.10 [d]	5.35 ± 1.1
Ash	4.76 ± 0.08 [d]	3.90 ± 0.02 [b]	4.12 ± 0.07 [c]	3.50 ± 0.04 [a]	3.40 ± 0.07 [a]	3.9 ± 0.5
Lipids	30.59 ± 0.04 [b]	32.78 ± 0.02 [d]	32.23 ± 0.05 [c]	29.31 ± 0.05 [a]	32.31 ± 0.06 [c]	31.44 ± 0.9
Protein	35.39 ± 0.05 [e]	30.08 ± 0.3 [a]	33.64 ± 0.06 [c]	35.68 ± 0.05 [d]	33.35 ± 0.05 [b]	33.63 ± 1.9
Calcium	80 ± 0 [a]	100 ± 0 [c]	80 ± 0 [a]	90 ± 0 [b]	80 ± 0 [a]	86 ± 0
Phosphorus	73 ± 5.7 [b]	73 ± 5.7 [b]	63 ± 5.7 [a]	110 ± 0 [c]	63 ± 5.7 [a]	76.6 ± 19.3
Sodium	20 ± 0 [a]	21 ± 0 [a]	22 ± 0 [a]	23 ± 0 [a]	24 ± 0 [a]	22 ± 0
Potassium	630 ± 0 [c]	690 ± 0 [d]	690 ± 0 [d]	480 ± 34.6 [b]	440 ± 0 [a]	586 ± 118
Magnesium	380 ± 0 [d]	330 ± 0 [b]	350 ± 0 [c]	310 ± 0 [a]	323 ± 11.5 [b]	338.6 ± 27.24
Iron	8.0 ± 0 [c]	7.6 ± 0.28 [c]	5.8 ± 0.28 [a]	6.5 ± 0 [b]	6.6 ± 0.28 [b]	6.09 ± 0.88
Coper	0.1 ± 0 [a]	0.36 ± 0.23 [b]	0.1 ± 0 [a]	0.1 ± 0 [a]	0.5 ± 0 [b]	0.23 ± 0.18
Zinc	6.0 ± 0 [c]	7.2 ± 0.28 [e]	5.2 ± 0.28 [a]	5.5 ± 0 [b]	6.5 ± 0 [d]	6.1 ± 0.79
Manganese	4.83 ± 0.57 [c]	1.16 ± 0.28 [a]	2.83 ± 0.28 [b]	2.5 ± 0 [b]	3.0 ± 0 [b]	2.8 ± 1.31

Different letters in the superscripts, between columns, indicate significant differences ($p \leq 0.05$).

High variability was observed in the copper, manganese, and phosphorus concentrations based on where the seeds were purchased. When compared with food composition tables and databases [5,6], there was a lower concentration of phosphorus, potassium, and magnesium in the seeds from Guatemala.

3.2. Macronutrients and Minerals in Pepita

Table 2 shows the macronutrient and mineral content of *pepita* from different places.

Table 2. Macronutrients (g/100 g) and minerals (mg/100 g) in *pepita*.

	Jutiapa	Flores, Petén	Santa Cruz, Quiché	Jacaltenango	Santiago Atitlán	Media
Water	4.45 ± 0.1 [b]	2.53 ± 0.08 [a]	4.69 ± 0.14 [b]	5.61 ± 0.28 [c]	5.97 ± 0.94 [c]	4.65 ± 1.39
Ash	9.43 ± 0.07 [d]	4.21 ± 0.03 [b]	9.35 ± 0.07 [d]	3.72 ± 0.03 [a]	4.52 ± 0.06 [c]	6.25 ± 2.88
Lipids	31.38 ± 0.06 [c]	32.65 ± 0.04 [d]	35.51 ± 0.04 [e]	28.32 ± 0.10 [b]	23.26 ± 0.11 [a]	30.22 ± 4.67
Protein	21.06 ± 0.02 [a]	32.61 ± 0.04 [d]	28.73 ± 0.04 [b]	33.46 ± 0.03 [e]	30.21 ± 0.03 [c]	29.21 ± 4.92
Calcium	100 ± 0 [b]	90 ± 0 [a]	120 ± 0 [c]	123 ± 5.7 [c]	250 ± 0 [d]	136 ± 64.8
Phosphorus	80 ± 0 [b]	73 ± 5.7 [b]	103 ± 5.7 [c]	63 ± 5.7 [a]	103 ± 5.7 [c]	84.6 ± 18.0
Sodium	2130 ± 0 [e]	20 ± 0 [a]	1320 ± 0 [d]	60 ± 0 [b]	880 ± 0 [c]	882 ± 889
Potassium	920 ± 34.64 [e]	690 ± 0 [c]	810 ± 0 [d]	650 ± 0 [b]	530 ± 0 [a]	720 ± 150
Magnesium	350 ± 0 [c]	380 ± 0 [d]	430 ± 0 [e]	280 ± 0 [b]	160 ± 0 [a]	320 ± 104.6
Iron	5.0 ± 0 [c]	6.66 ± 0.28 [d]	4.33 ± 0.28 [b]	2.0 ± 0 [a]	4.5 ± 0 [b]	4.5 ± 1.7
Coper	0.1 ± 0 [a]	0.23 ± 0.23 [a]	0.1 ± 0 [a]	0.1 ± 0 [a]	0.1 ± 0 [a]	0.12 ± 0.05
Zinc	3.5 ± 0 [c]	5.5 ± 0 [e]	1.5 ± 0 [a]	3.0 ± 0 [b]	5.2 ± 0.29 [d]	3.73 ± 1.64
Manganese	1.0 ± 0 [b]	3.0 ± 0 [c]	0.5 ± 0 [a]	3.5 ± 0 [d]	3.0 ± 0 [c]	2.2 ± 1.3

Different letters in the superscripts, between columns, indicate significant differences ($p < 0.05$).

It was observed that the *pepita* from Jutiapa and Santa Cruz del Quiché had twice the ash as those from other origins; it was also observed that the *pepita* from Flores, Petén, did not show variation in sodium content when compared to the other seeds, which can be interpreted as meaning that no salt was added. This is related to observations about the *pepita's* sodium content and other minerals, which increase from 50 to 100 times. Except for iron and copper, it was observed that all minerals had higher concentrations in the *pepita* than in the seed.

The bioavailability of minerals and protein in the pumpkin seed is expected to improve when the seed is ground into coarse flour, since the physical structure of the seed is broken and the external protective layers are disintegrated, which allows a greater surface area of contact between the tissues and digestive enzymes.

When comparing the average content of macronutrients and minerals in pumpkin seed and *pepita* (Table 3), the main differences were found in minerals, because the preparation of the *pepita* includes the addition of salt. Although the differences in macronutrient and mineral contents were not significant, the mathematical differences in sodium content were notable; since the sodium concentration was in the range of 20 to 2130 mg/100 g, this means a range of 1 to 166 mg of sodium per teaspoon of pumpkin seed.

Table 3. Media of macronutrients and minerals in *C. moschata* seeds and *pepita*.

	Seeds	*Pepita*
Water	5.35 ± 1.1	4.65 ± 1.39
Ash	3.9 ± 0.5	6.25 ± 2.88
Fat	31.44 ± 0.9	30.22 ± 4.67
Protein	33.63 ± 1.9	29.21 ± 4.92
Calcium	86 ± 0	136 ± 64.8
Phosphorus	76.6 ± 19.3	84.6 ± 18.0
Sodium	22 ± 0	882 ± 889
Potassium	586 ± 118	720 ± 150
Magnesium	338.6 ± 27.24	320 ± 104.6
Iron	6.09 ± 0.88	4.5 ± 1.7
Coper	0.23 ± 0.18	0.12 ± 0.05
Zinc	6.1 ± 0.79	3.73 ± 1.64
Manganese	2.8 ± 1.31	2.2 ± 1.3

4. Conclusions

Pumpkin seeds and *pepita* have nutritional characteristics that have been valued for thousands of years. Today, pumpkin seeds are consumed as an ingredient in sauces or an accompaniment to fruit. This study found differences in the macronutrient and mineral content of pumpkin seeds and *pepita*, depending on their origin. Its high protein, fat, and sodium content were also found, particularly in the *pepita*, due to its artisanal production.

Author Contributions: Conceptualization, methodology and original draft preparation E.J.S.; software, M.E.C. All authors have read and agreed to the published version of the manuscript.

Funding: This research received no external funding.

Institutional Review Board Statement: Not applicable.

Informed Consent Statement: Not applicable.

Data Availability Statement: The raw data supporting the conclusions of this article will be available from the authors upon request.

Acknowledgments: This work was supported by Universidad de San Carlos de Guatemala.

Conflicts of Interest: The authors declare no conflict of interest.

References

1. Lira Saade, R. Estudios Taxonómicos y Ecogeográficos de las Curcubitaceae Latinoamericanas de Importancia Económica. Instituto de Biología. UNAM, México e IPGRI. 294 p. 1995. Available online: https://www.researchgate.net/publication/48214811_Estudios_taxonomicos_y_ecogeograficos_de_las_Cucurbitaceae_Latinoamericanas_de_importancia_economica (accessed on 26 August 2024).
2. Vestal, P.A. Cucurbita Moschata Encontrada en Montículos Precolombinos en Guatemala. *Harv. Univ. Bot. Leafl.* **1938**, *6*, 65–69. Available online: https://www.biodiversitylibrary.org/part/168392 (accessed on 29 August 2024).
3. Sagastume, E. Chereban sakil. El iwaxte de Santa Catarina Barahona, Sacatepéquez. Guatemala. *La Tradic. Pop.* **2023**, *264*, 1–20. Available online: https://revistasceceg.usac.edu.gt/detallearticulo.php?c=556&a=2106 (accessed on 26 August 2024).
4. Fennema, O.R. Química de los Alimentos. Food Chemistry. Tercera Edición. Editorial Acribia, Zaragoza, Espa. 2000. Available online: https://sceqa.wordpress.com/wp-content/uploads/2014/05/quc3admica-de-los-alimentos-fennema.pdf (accessed on 27 August 2024).
5. Departamento de Agricultura de los Estados Unidos. USDA Database. Consultado el 7 de agosto 2024. Available online: https://fdc.nal.usda.gov/fdc-app.html#/food-details/2515380/nutrients (accessed on 29 August 2024).

6. Instituto de Nutrición de Centroamérica y Panamá. *Tabla de Composición de Alimentos de Centroamérica*, 2nd ed.; Menchú, M.T., y Méndez, H., Eds.; INCAP: Cdad. de Guatemala, Guatemala, 2012. Available online: https://www.sennutricion.org/media/tablas/INCAP.pdf (accessed on 26 August 2024).
7. Alfawaz, M.A. Chemical Composition and Oil Characteristics of Pumpkin (*Cucurbita maxima*) Seed Kernels. *Res. Bult.* **2004**, *129*, 5–18. Available online: https://www.researchgate.net/publication/237427083_Chemical_Composition_and_Oil_Characteristics_of_Pumpkin_Cucurbita_maxima_Seed_Kernels (accessed on 28 August 2024).
8. Devi, N.M.; Prasad, R.V.; Palmei, G. Caracterización fisicoquímica de semillas de calabaza. *Int. J. Chem. Stud.* **2018**, *6*, 828–831. Available online: https://www.researchgate.net/publication/327751786_Physico-chemical_characterisation_of_pumpkin_seeds (accessed on 27 August 2024).
9. Petkova, Z.Y.; y Antova, G.A. Changes in the composition of pumpkin seeds (*Cucurbita moschata*) during development and maturation. *Grasas Y Aceites* **2015**, *66*, e058. [CrossRef]

biology and life sciences forum

MDPI

Proceeding Paper

Technological Development of an Instant Product Based on Fermented Purple Corn (*Zea mays* L.) Beverage †

José Meléndez [1], Oscar Malpartida [1,*] and Nancy Ascención Chasquibol [2]

[1] Carrera de Ingeniería Industrial, Universidad de Lima, Av. Javier Prado Este 4600, Surco, Lima 15023, Peru; 20152075@aloe.ulima.edu.pe

[2] Ingeniería Industrial, Grupo de Investigación en Alimentos Funcionales, Instituto de Investigación Científica, Universidad de Lima, Av. Javier Prado Este 4600, Fundo Monterrico Chico, Surco, Lima 15023, Peru; nchasquibol@ulima.edu.pe

* Correspondence: 20092410@aloe.ulima.edu.pe

† Presented at the VI International Congress la ValSe-Food, Lima, Peru, 23–25 September 2024.

Abstract: Chicha de güiñapo (ChG) is an ancestral beverage from the culture and gastronomy of Arequipa, Peru. This traditional drink is made from purple corn (*Zea mays* L.), cultivated across various Peruvian regions. Purple corn is renowned for its nutritional content and high bioactive compound value, such as antioxidants (20.5 ± 2.0 μmol TE/g), total phenolic compounds (2.5 ± 0.3 mg GAE/g), and anthocyanins (1.8 ± 0.2 mg/g). This research aimed to explore the technological development of an instant powder product derived from chicha de güiñapo (ChG) utilizing spray-drying technology. The purple corn (*Zea mays* L.) used in this study was from Peru; it was first processed by boiling the güiñapo at 100 °C 1 h, followed by cooling and fermenting under controlled conditions for 5–7 days until achieving the desired characteristics referenced from previous studies, such as pH, alcohol content (v/v), and degrees Brix. Upon attaining the desired fermentation characteristics, the ChG was centrifuged, filtered, and dehydrated by spray-drying technology with the following parameters: air inlet temperature (165 °C), airflow (0.89 mL/min), feed flow (1.67 mL/min), and outlet temperature (93 °C). These optimal parameters were determined using the response surface methodology after 15 runs. Then, a fine purple powder was produced with 6.61% moisture, pH 4.83, and 1.5 °Brix. The results of proximal analysis before and after spray-drying were for carbohydrates (1.77% to 82.67%), ash (0.02% to 4.91%), protein (0.10% to 5.81%), and alcohol (3.17% to 0.64%). This study highlights the biodiversity, sustainability, and food security of ancestral crops to contribute to cultural heritage valorization.

Keywords: Andean beverages; chicha de güiñapo; fermentation; purple corn; spray-drying; traditional food preservation

1. Introduction

Chicha de güiñapo (ChG) is an ancestral drink deeply rooted in the culture and gastronomy of Arequipa, Peru. Made from purple corn (*Zea mays* L.) grown in several Peruvian regions, this traditional drink is recognized for its high nutritional content and bioactive properties [1].

The main raw material to make ChG is purple corn (*Zea mays* L.), which is partially germinated and then ground to produce güiñapo, a key ingredient in the traditional preparation of ChG. Purple corn is grown in several regions of Peru, with approximately 5000–6000 hectares dedicated to its production. This crop, originally from the Andean region, grows in environments up to 3000 m.a.s.l. and is known for its high content of anthocyanins, such as cyanidin-3-glucoside, characteristic of its purple color [2]. The use of güiñapo in ChG, particularly in the Arequipa region, is attributed to tradition, making it unique compared to chicha from other regions of Peru [1]. Arequipa, located at an average altitude of 2335 m, is known for its unique blend of Andean and Spanish culinary traditions, with ChG being singled out for its use of partially germinated purple corn, a practice preserved since the pre-Incan period [1].

Spray-drying, a technique that turns liquid food products into powder, helps preserve nutritional and bioactive properties, ensuring product stability and quality [3,4].

This research aimed to convert ChG into powder using spray-drying technology.

2. Materials and Methods

2.1. Raw Materials and Preparation of Chicha de Güiñapo

The raw material used was purple corn (*Zea mays* L.), which is traditionally cultivated in various regions of Peru, particularly in the Andean highlands, and it was purchased at the San Camilo market located in the city of Arequipa. This corn is partially germinated and then milled to produce germinated güiñapo. The purple corn (*Zea mays* L.), originating from the rich agricultural traditions of several Peruvian regions, used in this study was processed by boiling the güiñapo at 100 °C for one hour, followed by cooling and fermentation under controlled conditions for 5–7 days until reaching the desired characteristics referenced in previous studies, such as pH, alcohol content (*v*/*v*), and degrees Brix [1]. The fermentation process is critical in developing the distinctive taste and bioactive properties of the ChG. After reaching the desired fermentation characteristics, the ChG is centrifuged and filtered to remove solid residues, preparing it for the subsequent spray-drying process. Fermentation occurs naturally, based on the native microorganisms present in the environment and in the raw materials. No external inoculum is added, which allows the process to conserve the traditional methods and flavors of the region of Arequipa [4].

2.2. Spray-Drying Process

The spray-drying process was employed to convert ChG into a fine powder, optimizing key parameters to maintain the beverage's nutritional and functional qualities. The ChG was first prepared as described previously, then dried via spray-drying (Büchi B-290, Büchi Labortechnik AG, Flawil, Switzerland). The optimal conditions were determined through the response surface methodology with 15 runs in Minitab 19 software (Stat-Ease, Inc., Minneapolis, MN, USA) [3]. Three independent variables were used: X1—temperature (135–165 °C); X2—air flow (0.85–0.95 mL/min); and X3—feed flow (0.1–0.2 mL/min). Water activity was measured using a water-in-food activity meter, (WA-60A, Guangzhou Landtek Instruments Co., Guangzhou, China), and the result was expressed for each run. The yield was calculated with the weight of the solution before entering the spray-drying process divided by the weight of the resulting solution after the drying process and expressed as a percentage. Alcohol was measured in terms of variables of response (Table 1).

Table 1. Response surface methodology with 15 runs.

StdOrder	RunOrder	PtType	Blocks	Temperature (°C)	Air Flow (mL/min)	Feed Flow (mL/min)	Moisture (%)	Water Activity (WA-60A)	Alcohol (%)	Yield (%)
6	1	2	1	165	0.9	0.1	6.1 ± 1.02	0.44 ± 1.10	0.5 ± 1.1	0.96 ± 1.10
15	2	0	1	150	0.9	0.15	8.54 ± 0.98	0.52 ± 0.75	0.5 ± 1.20	0.57 ± 1.05
13	3	0	1	150	0.9	0.15	8.17 ± 0.76	0.51 ± 0.70	0.5 ± 1.5	0.55 ± 1.05
4	4	2	1	165	0.95	0.15	7.71 ± 0.43	0.56 ± 0.80	2 ± 0.95	0.58 ± 1.15
5	5	2	1	135	0.9	0.1	6.68 ± 0.78	0.52 ± 0.79	1.5 ± 0.9	0.57 ± 1.20
1	6	2	1	135	0.85	0.15	6.23 ± 0.97	0.46 ± 1.10	1 ± 0.85	0.89 ± 1.05
3	7	2	1	135	0.95	0.15	6.75 ± 0.90	0.45 ± 1.20	1.5 ± 1.15	0.84 ± 1.10
2	8	2	1	165	0.85	0.15	10.53 ± 0.70	0.53 ± 1.15	2.5 ± 1.25	0.70 ± 1.15
9	9	2	1	150	0.85	0.1	6.96 ± 1.01	0.53 ± 0.70	3.5 ± 0.85	0.49 ± 0.90
7	10	2	1	135	0.9	0.2	11.61 ± 0.900	0.47 ± 1.25	0.5 ± 0.95	0.42 ± 0.85
8	11	2	1	165	0.9	0.2	6.61± 0.80	0.52 ± 1.05	0.5 ± 1.10	0.30 ± 1.15
14	12	0	1	150	0.9	0.15	9.6 ± 1.10	0.87 ± 1.05	1.5 ± 1.15	0.26 ± 0.95
12	13	2	1	150	0.95	0.2	8.07 ± 1.20	0.87 ± 0.80	2 ± 1.0	0.23 ± 1.15
11	14	2	1	150	0.85	0.2	11.63 ± 1.3	0.73 ± 0.90	2 ± 0.85	0.13 ± 0.95
10	15	2	1	150	0.95	0.1	10.85 ± 1.0	0.62 ± 1.30	0.5 ± 1.01	0.36 ± 0.80

2.3. Proximal Analysis

The proximal composition analysis was carried out according to official methods [4]. The moisture content was determined at 110 °C until constant weight using a halogen moisture analyzer (Sartorius MA-30, Sartorius AG, Göttingen, Germany). The total protein content was determined as % nitrogen × 6.25 using a Kjeldahl analyzer (UDK 139, VELP, Usmate Velate, Italy). The ash content was determined by incineration at 550 °C for 72 h in a muffle furnace. The fat content was determined by extraction with hexane for 4 h, as described in [4].

2.4. Total Phenolic Content

The total phenolic content (TPC) was determined by the Folin–Ciocalteau method [4] at 760 nm using a spectrophotometer (UV 1280 Vis Spectrophotometer Shimadzu, Kyoto, Japan). The results were expressed as µg of gallic acid equivalent (GAE)/g powder. Analyses were performed in triplicate and are presented as mean values.

2.5. Antioxidant Activity

The antioxidant activity was determined by the DPPH method [4] with some modifications at 517 nm by spectrometry (UV 1280 Vis Spectrophotometer Shimadzu, Kyoto, Japan). The results were expressed as µg Trolox/g sample. Analyzes were performed in triplicate and presented as mean values.

2.6. Analysis of pH, °Brix, and Alcohol Content

The pH of the samples was measured using a digital pH meter (Metrohm 827 pH Lab, Metrohm AG, Herisau, Switzerland). The soluble solid content (°Brix) was measured with a portable digital refractometer (Atago PAL-1, Atago Co., Tokyo, Japan). The alcohol content was initially measured using a portable digital refractometer (Atago PAL-34S, Atago Co., Tokyo, Japan) with a measurement range of 0 to 45 g/100 g, and the results were then converted to percentage volume/volume (v/v) based on the density of the solution. These measurements were conducted to monitor the fermentation parameters of ChG.

2.7. Statistical Analysis

Results were expressed as mean ± standard deviation. All measurements were determined in duplicate or triplicate. Analysis of variance (ANOVA) followed by a post hoc test was used to determine the statistical significance of differences between sample means. It was used to analyze the data acquired at a 95% significance level with Minitab 19.0 Software (Minitab Inc., State College, Palo Alto, CA, USA).

3. Results and Discussion

3.1. Spray-Drying

The optimum conditions for the ChG powder drying process were predicted using the response surface methodology with a Box–Behnken design (Table 1). The p-value of the coefficients related to the variables were as follows: Temperature (°C), Air flow (mL/min), and Feed flow mL/min. All terms resulted in a p-value less than or equal to $\alpha = 0.05$, so there was an association between the response variable and the terms presented. Regression statistics, such as coefficient of determination R^2 (94.01%), adjusted coefficient of determination R^2_{adj} (83.24%), and predicted coefficient of determination R^2_{pred} (78.42%), would indicate an acceptable degree of evaluation between the model and the response variable.

The optimization of the ChG drying process through the second-order quadratic model is presented in Figure 1. The optimal conditions of the Box–Behnken design were Temperature (165 °C), Air flow (0.89 mL/min), and Feed flow (1.67 mL/min) with 6.61% moisture, 0.52 water activity, 0.50% alcohol, and 0.30% yield.

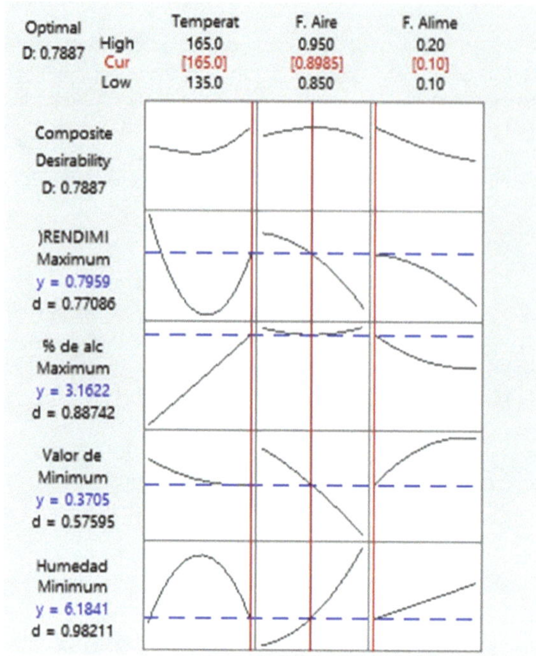

Figure 1. Optimization of the Box–Behnken design obtained using Minitab 19 software.

The shapes of the contour curves can be seen in Figure 2. The shapes of the curves indicate the importance of the interactions between the variables studied (temperature, air flow, and feed flow) with the response variables.

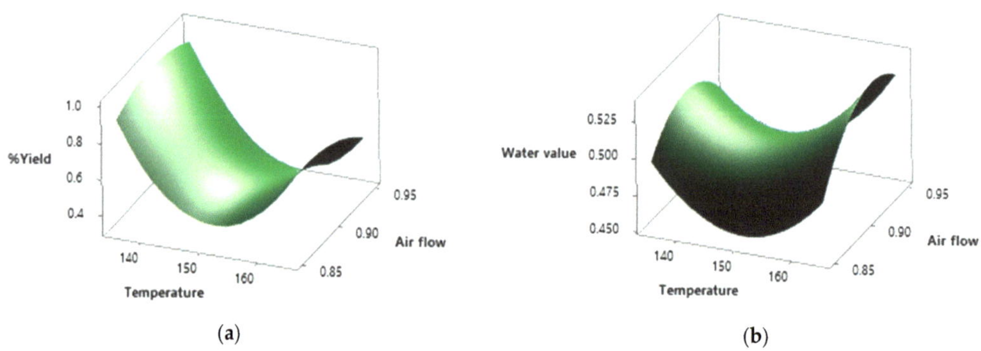

Figure 2. Surface plot of the optimized ChG drying process. (**a**) Surface graph of the efficiency (%) of the optimized ChG drying process as a function of temperature (°C) and air flow (mL/min). (**b**) Surface graph of the water value of the optimized ChG drying process as a function of temperature (°C) and air flow (mL/min).

3.2. Characterization

Table 2 shows the results of pH, °Brix, and % of alcohol of ChG development. Under optimal conditions, the pH (5.08 ± 0.13) of the ChG development was higher than that of the commercial ChG (3.22 ± 0.11). The °Brix (1.97 ± 0.50) and % of alcohol (2.5 ± 1.00%) of ChG development were lower than the those of the commercial ChG (7.27 ± 0.50 °Brix and

13 ± 3.60% alcohol). This is because in the commercial ChG, sugars such as sucrose are added to increase the °Brix and % of alcohol [4].

Table 2. Parameters physicochemicals obtained after the fermentation of chicha de güiñapo (ChG).

Parameter	Commercial ChG	Development ChG
pH	3.22 ± 0.11 ^{b}	5.08 ± 0.13 ^{a}
°Brix	7.27 ± 0.50 ^{a}	1.97 ± 0.50 ^{b}
Alcohol (%)	13 ± 3.60 ^{a}	2.5 ± 1.00 ^{b}

Results are expressed as means ± SD (n = 3). a, b values in the same row with different letters differ significantly when $p < 0.05$.

The proximal analysis of purple Corn, *güiñapo* flour, commercial ChG, and powdered ChG is shown in Table 3.

Table 3. Proximal analysis of purple corn, güiñapo flour, commercial ChG, and powdered ChG.

Component (%)	Purple Corn (*Zea mays*)	*Güiñapo* Flour	Commercial ChG	Development ChG	Powder ChG
Moisture	8.03 ± 1.05 ^{d}	14.31 ± 0.23 ^{c}	89.82 ± 0.32 ^{b}	98.11 ± 0.35 ^{a}	8.29 ± 2.03 ^{d}
Carbohydrates	76.56 ± 1 ^{*a}	68.94 ± 1 $^{\#a}$	10.12 ± 1 $^{\#b}$	1.77 ± 1 ^{*b}	11.62 ± 0.3 ^{*b}
Protein	2.59 ± 0.51 ^{*c}	11.72 ± 0.23 $^{\#a}$	0.05 ± 0.03 $^{\#d}$	0.1 ± 0.05 ^{*d}	5.81 ± 0.33 ^{*b}
Ash	10.38 ± 0.02 ^{a}	2.64 ± 0.08 ^{c}	0.01 ± 0.01 ^{d}	0.02 ± 0.02 ^{d}	4.91 ± 0.03 ^{b}

* (g/100g). # (g/100mL). Results are expressed as means ± SD (n = 3). a, b, c, d values in the same column with different letters differ significantly when $p < 0.05$.

The total phenolic and antioxidant content (DPPH) of *güiñapo* flour, commercial ChG, and development ChG are shown in Table 4. The spray-dried powdered ChG showed a higher antioxidant capacity (7743.454 ± 48.767 mg Trolox/100 g d.m) and a high phenolic content (0.700 ± 0.194 mg GAE/100 g d.m). The *güiñapo* flour presented lower antioxidant (9502.90 ± 14.66 mg Trolox/100 g d.m) and phenolic content (0.085 ± 0.070 mg GAE/100 g d.m).

Table 4. Total phenolic content and antioxidant activity (DPPH) in *güiñapo* flour, commercial ChG, and development ChG.

Type of Analysis	DPPH (mg Trolox/100 g dm)	SPC (mg GAE/100 g dm)
Powdered ChG	7743.454 ± 48.767 ^{b}	0.700 ± 0.194 ^{b}
Güiñapo flour	9502.90 ± 14.66 ^{a}	0.085 ± 0.070 ^{a}
Commercial ChG	0.185 ± 0.001 ^{c}	0.00 ± 0.00 ^{a}
Development ChG	0.198 ± 0.004 ^{c}	0.00 ± 0.0 ^{a}

Results are expressed as means ± SD (n = 3). a, b, c values in the same column with different letters differ significantly when $p < 0.05$.

The commercial ChG showed a higher antioxidant capacity (0.185 ± 0.001 mg Trolox/100 g d.m) and a high phenolic content (0.00 ± 0.00 mgGAE/100 g d.m) compared to the development ChG antioxidant capacity (0.198 ± 0.004 mg Trolox/100 g d.m) and phelonic content (0.00 ± 0.00 mg GAE/100 g d.m).

These results suggest that the spray-drying process, under optimized conditions, effectively preserves the bioactive compounds in the final powder product.

4. Conclusions

This research demonstrates the potential of using spray-drying technology to develop a stable, nutritionally rich, and culturally significant instant powder product from *güiñapo* ChG. The optimized spray-drying parameters ensured high retention of essential nutrients and bioactive compounds; the process not only preserved the traditional qualities of *güiñapo* ChG but also enhanced its potential for broader consumption, contributing to the valorization of cultural heritage and the promotion of ancestral crops.

Author Contributions: Conceptualization, J.M., O.M. and N.A.C.; methodology, J.M., O.M. and N.A.C.; software, N.A.C.; formal analysis, J.M. and O.M.; investigation, J.M., O.M. and N.A.C.; resources, N.A.C.; data curation, J.M. and O.M.; writing—original draft preparation, J.M. and O.M.; supervision, N.A.C. All authors have read and agreed to the published version of the manuscript.

Funding: This research received no external funding.

Institutional Review Board Statement: Not applicable.

Informed Consent Statement: Not applicable.

Data Availability Statement: Data is contained within the manuscript.

Acknowledgments: This work was supported and developed at Laboratorio de Alimentos Funcionales de la Carrera de Ingeniería Industrial, Universidad de Lima, Peru.

Conflicts of Interest: The authors declare no conflicts of interest.

References

1. Vargas, D.; Aguilar, B.; Pezo, N. Ancestral Peruvian ethnic fermented beverage "Chicha" based on purple corn (*Zea mays* L.): Unraveling the health-relevant functional benefits. *J. Ethn. Foods* **2020**, *7*, 35. [CrossRef]
2. Medina, A.; Narro, L.; Chávez, A. Purple corn (*Zea mays* L.) crop in the Peruvian Highlands: Adaptation and identification of high-yield and high anthocyanin content cultivars. *Agric. Res.* **2017**, *56*, 291–299. [CrossRef]
3. Gül, O.; Atalar, İ.; Törnük, F.; Akgün, A. Process optimization of a cereal-based fermented beverage (Boza) powder and investigating upscaling conditions. *Food Bioprod. Process.* **2018**, *96*, e14248. [CrossRef]
4. Chasquibol, N.; Sotelo, A.; Alarcón, R. Development of Powdered Beverage with Cushuro (*Nostoc commune*) Concentrated Protein and Quinoa (*Chenopodium quinoa*). *Biol. Life Sci. Forum* **2023**, *25*, 2. [CrossRef]

biology and life sciences forum

MDPI

Proceeding Paper

Compositional Changes Associated with Successive Boiling of Wild *Cynophalla retusa* (Indian Bean) Pods Collected from the Paraguayan Chaco †

Adecia Suarez, Laura Mereles *, Patricia Piris, Rocio Villalba, Olga Heinichen and Silvia Caballero

Dirección de Investigación, Facultad de Ciencias Químicas, Universidad Nacional de Asunción, San Lorenzo P.O. Box 1055, Paraguay; adecia.suarez@gmail.com (A.S.); ppiris@qui.una.py (P.P.); rvillalba@qui.una.py (R.V.); olgahena@qui.una.py (O.H.); scaballero@qui.una.py (S.C.)

* Correspondence: laurameereles@qui.una.py

† Presented at the VI International Congress la ValSe-Food, Lima, Peru, 23–25 September 2024.

Abstract: *Cynophalla retusa*, known as "Indian bean", is an important traditional food for the ethnic groups of the Gran Chaco. However, its contribution of minerals to the diet is unknown and the toxic nature of its raw pods has been reported. The aim of this investigation was to evaluate the composition of minerals, oxalic acid and phytate contents in whole raw and cooked pods, with successive changes of boiling water every 1 h for 4 h in total, as well as the alkaloid content in the cooking water. Bivalent mineral composition determinations (Ca, Fe, Cu and Mg) were made, as well as measurements of the phosphorus and antinutrient contents, such as phytate and oxalic acid, to determine the mineral contribution. The raw pods (*C. retusa*) contained 6.67% ash, with high contents of Ca, Fe, Cu, Mg and P. Loss of minerals occurred with successive boiling and significant decreases in antinutrients, with significant changes after each boiling period (1, 2, 3, 4 h). The boiling improved the bioavailability of Ca by removing oxalic acid from the sample cooked in the fourth boiling period. However, the phytate contents were not reduced to the same extent (only up to 40%). The results show that *C. retusa* pods can be a source of minerals (Ca, Fe, Cu and Mg) under controlled conditions of cooking and decreases in antinutrients like oxalic acid. From this perspective, this food source can be a viable alternative to increase food safety and nutrition, using one of many Paraguayan species that are little-known. Therefore, domestication and conservation studies are necessary.

Keywords: antinutrients; *Cynophalla retusa*; oxalic acid; phytate; neglected and underutilized species; minerals

1. Introduction

Regional food resources have great value within the framework of food security, and there are many indigenous foods that have been underutilized until now. However, they constitute means of subsistence in indigenous populations of the Central Chaco, as is the case with the "Indian bean", "sacha bean" and "guaicuru bean" (*Cynophalla retusa* (Griseb.) X. Cornejo and H.H. Iltis. This name was recently accepted name for the flora of the Southern Cone, and it is synonymous with *Capparis retusa* Griseb., *Capparis retusa* Griseb. var. velutina, *Capparis cynophallophora* L. var. *cuneata*, *Capparis cynophallophora* L. var. *retusa*). The genus *Cynophalla* (DC.) J. Presl (Capparaceae) comprises 16 woody species, was established to group a series of species previously found in the genus *Capparis* L. [1]. One of the main limitations for its integral use is the lack of knowledge about its real nutritional value and its wild appearance, so studies on its nutritional potential in ancestral consumption conditions and its adaptation to standardized cultivation systems can be a way to improve the nutrition of the population and generate market opportunities in the productive sector of non-traditional crops.

Cynophalla retusa it is a shrub typical of the Gran Chaco ecosystem, 2–7 m high, and with a silicuiform, incurved capsular fruit, with notable strangulations, whose ripening season occurs from December to March. The plant is considered toxic and its fruit, although edible, is cooked in a particular way known by the local communities of the indigenous peoples of the Gran Chaco, such as the Lengua–Maskoy, Tobas, Wichis, Qom, Qomle'ec and Pilagá. Ethnobotanical studies report that, in human food, the boiled fruits of *Cynophalla retusa* are consumed with successive changes of water, in order to extract their "bitter beginning" [2,3]. It has been observed that *C. retusa* raw grains contain antinutrients, such as oxalic acid (87.8–121.0 mg/100 g) and phytate (463.6–535.7 mg/100 g), which can interfere with the bioavailability of minerals [4].

Although plant foods can be important sources of minerals in the diet, they can have the presence of oxalic acid and phytates (*myo*-inositol hexakisphosphate; InsP_6), two important components that can affect the bioavailability of bivalent minerals, because they potentially form complexes with dietary minerals and proteins, such as phytates, or by forming salts with oxalic acid, and decrease their bioavailability. These compounds can be reduced by boiling and solubilization in cooking water, or by treatments such as pre-soaking, germination or fermentation [5]. The choice of method for the reduction in these undesirable compounds depends largely on the type of food and the form of the final product in which that food is consumed. The aim of this investigation was to evaluate the composition of mineral, oxalic acid and phytate contents in whole raw and cooked *Cynophalla retusa* pods, with successive changes to the water in which the *C. retusa* is boiled.

2. Materials and Methods

2.1. Raw Material

The samples used for the study were the closed mature capsules of *Cynophalla retusa* from Philadelphia, Chaco, Paraguay, collected in 2021. Once in the Food Biochemistry department laboratory (FCQ-UNA), they were preserved at −20 °C until the time of analysis.

2.2. Experimental Design

A triplicate experiment was performed at a laboratory scale, according to the cooking process reported in the literature [3] for the raw mature whole capsules of *Cynophalla retusa*, with successive changes of water every 1 h; here, the capsules were boiled for 4 h in total (Figure 1). Bivalent mineral composition determinations (Ca, Fe, Cu and Mg) were made, as well as measurements of phosphorus and antinutrient contents, such as phytate and oxalic acid, to determine the mineral contribution in the cooking samples. The presence of alkaloids was also determined in the cooking water. The mineral, oxalic acid and phytate contents in whole raw and cooked *C. retusa* pods were evaluated.

Figure 1. *Cont.*

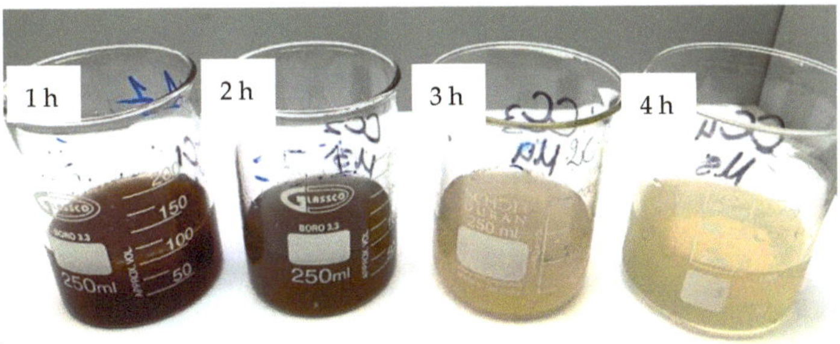

(d)

Figure 1. Process of sample cooking: (**a**) *Cynophalla retusa* fresh whole pods. (**b**) Pods boiled with water. (**c**) Samples drained after the first boiling period (1 h). (**d**) Appearance of the cooking broth after water changes at 1 h, 2 h, 3 h and 4 h.

2.3. Analysis

Minerals (Ca, Fe, Mg, Cu) were determined by official AOAC 968.08 method [6] by atomic absorption spectrophotometry (AAS) in a SHIMADZU model AA 6300 equipment (Kyoto, Japan). Results were expressed in mg/100 g. The phosphorus content was analyzed by the AOAC 970.39 method, measuring the percentage of transmittance, at a wavelength of 400 nm UV, on a SHIMADZU device, model UV-1800 (Kyoto, Japan). Results were expressed in mg/100 g. The phytate content was determined by the method previously described by García et al. (1982), by complexometric titration of the excess Fe (III) with EDTA as the titrating chelating agent and 5-sulfosalicylic acid as an indicator. Results were expressed in phytic acid equivalents (PAE/100 g). The determination of oxalic acid was performed according to the method described by Dona and Vercheret [7], by spectrophotometry at 305 nm, due to dissociation of the zirconia (IV)–chloranilate complex.

3. Results and Discussion

Mineral Composition

The raw pods (*C. retusa*) contained 6.67% ash, with high contents of Ca, Fe, Cu, Mg and P. According to the contents of the bivalent minerals (Ca, Fe, Cu and Mg), a significant decrease could be observed after 1 h of cooking (Table 1). Variable values were observed in each cooking hour in water. However, at the end of the experiment, all the mineral concentrations were significantly lower in the cooked samples. Data on the mineral composition of this food are limited in the literature; to our knowledge, this is the first work on the effect of cooking on the mineral content of *Cynophalla retusa* capsules. Taking into account that the species *Cynophalla retusa* is considered synonymous with *Capparis retusa*, comparisons were made with other species of the *Capparis* genus. Thus, in the raw samples of *C. retusa*, the Ca content was comparable to that of *Capparis spinosa* "capers" (21.8–154 mg/100 g), and much higher compared to what has been reported in fruits of *Capparis decidua* (14.1–35.1 mg/100 g). The Fe contents reported for other Capparis fruits, such as *C. ovata* (5.2–43.9 mg/100 g), *C. spinosa* (6.9–25.4 mg/100 g) and *C. decidua* (12.3–81.8 mg/100 g) are higher than those observed in this work (1.57 $\pm$ 0.25 mg/100 g) [8].

Table 1. Contents of minerals, phytic acid and oxalic acid and presence of alkaloids in the raw and cooked samples.

Compound	Cooking Time in Water at 100 °C (h)				
	0 (raw)	1	2	3	4
Ca (mg/100 g)	87.1 ± 6.8 ᵃ	33.7 ± 5.3 ᵇ	38.1 ± 6.1 ᵇ	50.8 ± 2.3 ᶜ	39.1 ± 6.3 ᵇᶜ
Fe (mg/100 g)	1.57 ± 0.25 ᵃ	0.55 ± 0.19 ᵇ	0.370 ± 0.050 ᵇᶜ	0.34 ± 0.09 ᵇᶜ	1.06 ± 0.64 ᵇᶜ
Cu (mg/100 g)	0.73 ± 0.06 ᵃ	0.600 ± 0.028 ᵇ	0.51 ± 0.04 ᶜ	0.64 ± 0.01 ᵇᵈ	0.71 ± 0.03 ᵃ
Mg (mg/100 g)	48.3 ± 4.9 ᵃ	5.90 ± 0.78 ᵇ	8.78 ± 1.01 ᵇᶜ	11.9 ± 0.7 ᶜ	7.26 ± 1.13 ᵇᶜ
P (mg/100 g)	98.1 ± 4.2 ᵃ	24.3 ± 0.94 ᵇ	19.4 ± 0.7 ᵇᶜ	17.1 ± 0.7 ᶜ	14.4 ± 0.9 ᶜ
PA (PAE/100 g).	1950 ± 316 ᵃᵇ	1183 ± 365 ᵇ	989 ± 126 ᵇ	908 ± 25 ᵇ	790 ± 180 ᵇ
OA (mg/100 g).	191.1 ± 1.4 ᵃ	34.16 ± 6.83 ᵇ	25.6 ± 1.5 ᵇᶜ	21.1 ± 5.1	Not detectable
Alkaloids	+++	-	-	-	-

Note: The results are expressed as average ± SD. Different letters in the same row demonstrate statistically significant differences between means (single-factor ANOVA, post-Tukey test; $p < 0.05$). PAE: phytic acid (PA) equivalent, OA: oxalic acid. (+++) abundant presence, (-) absence/not detected.

However, previous data reported [4] indicate that in general, the grains of *C. retusa* have a potential contribution of magnesium (160–165 mg/100 g), iron (3.70–4.40 mg/100 g) and phosphorus (294–305 mg/100 g), with concentrations higher than those observed in the whole pods in this work. The P content in the raw samples (98.1 ± 4.2 mg/100 g) (Table 1) also decreased significantly in the first 1 h cooking period, with values much lower than those of the fruits of species such as *C. ovata* and *C. decidua* (290–3073 and 701–808 mg/100 g, respectively) [8]. The concentration of phytate decreased after 1 h of cooking, but then remained stable (Table 1). All the phytate values were higher than those reported in the grains of *C. retusa* 463.6–535.7 mg/100 g [4], which could indicate that the phytate content is higher in the pod that surrounds the grains in the fruit capsule. The oxalic acid content in the crude samples (191.1 ± 1.4 mg/100 g) was also higher than that reported for grains of *C. retusa* (87.8–121.0 mg/100 g) [4], and decreased from its initial value in the raw samples to levels not detectable after 4 h of cooking, with successive water changes. These results show that oxalic acid is lost in the cooking broth. However, phytates are more stable in the matrix and remain even with a boiling treatment of 4 h. The content of P found in the form of phytate was calculated to determine the amount of bioavailable free phosphorus in the samples, and the percentage of phosphorus as phytic acid is shown in Table 2. It is known that the limit value for the molar ratio of AF/Ca = 0.2, and that values above this would mean that PA compromises the good absorption of Ca [9]. On the other hand, a PA/Fe molar ratio > 0.4 also compromises the good absorption of Fe [10]. The results obtained allow us to understand that the bioavailability of Ca and Fe in both raw and cooked samples can be affected by phytate content (Table 2).

Table 2. Phytic acid and molar ratio between the antinutrients and minerals Ca and Fe.

Relation	Boiling Time (h)				
	0 (raw)	1	2	3	4
% P as PA	5.61 ± 0.94 ᵃ	13.7 ± 4.4 ᵇ	14.4 ± 2.2 ᵇ	14.9 ± 4.0 ᵇᶜ	15.6 ±4.5 ᶜ
PA/Ca	1.30 ± 0.23 ᵃ	1.20 ± 0.35 ᵃ	1.23 ± 0.15 ᵃ	1.00 ± 0.29 ᵃᵇ	0.75 ± 0.09 ᵇ
PA/Fe	116 ± 15 ᵃᵇ	206 ± 28 ᶜ	257 ± 38 ᶜᵈ	396 ± 298 ᵃᵈ	79 ± 26 ᵃ
OA/Ca	0.84 ± 0.04 ᶜ	0.23 ± 0.05 ᵇ	0.21 ± 0.01 ᵇ	0.15 ± 0.02 ᵃ	-

Note: The results are expressed as average ± standard deviation. Different letters in the same row demonstrate statistically significant differences between means (single-factor ANOVA, post-Tukey test; $p < 0.05$). % P as PA indicates the percentage of phosphorus as phytic acid. PA/Ca: phytic acid/Ca molar ratio. PA/Fe: phytic acid/Fe molar ratio. OA/Ca: oxalic acid/Ca molar ratio. (-) absence/not detected.

This can be explained in part by the fact that phytates are thermostable, and therefore a significant reduction in their content cannot be expected during the cooking process. A Phytate has traditionally been considered an antinutrient in animal feed and human diets, due to its potent chelating capacity. However, more recently, phytate has been recognized as a natural antioxidant and as a nutraceutical, and classified by the Food and Drug Administration (FDA) as Generally Recognized As Safe (GRAS), it has potent antioxidation and anti-inflammatory actions and it has been shown to be effective in treating or preventing certain diseases. Recommending a diet high in phytate can exert multiple health benefits, with no harm [9].

On the other hand, the ability of oxalic acid to interact with Ca, being a dicarboxylic acid that forms insoluble salts with calcium, produces a decrease in the bioavailability of this metal if the AO/Ca ratio > 1. It is considered that 2.5 g of oxalic acid precipitates with 1 g of calcium. Therefore, the bioavailability of calcium is determined by the oxalic acid/calcium molar ratio, and when this ratio is greater than 2.25, the consumption of the food can be decalcifying, since the calcium of the observed food not only complexes but, concomitantly, binds to calcium from other foods in the digestive tract, as well as precipitating Ca ions present in the intestinal lumen [9]. The calculation of the AO/Ca ratio made it possible to demonstrate that the water cooking treatment and the successive water changes were efficient for the removal of oxalic acid from the crude sample and improved the bioavailability of the calcium (Table 2). From this perspective, more studies on the optimization of cooking treatments are necessary to achieve the nutritionally efficient use of the minerals present in this native food resource, such as with the previous use of phytase enzymes, without ruling out that this food source can be a viable alternative to increase food security and nutrition, in local communities in situations of scarcity, following the traditional procedures validated here. It is necessary to design conservation programs and use plans to improve the sustainability of food systems in which cooking methods are used in an ancestral way.

4. Conclusions

In the sample of the raw whole fruit, a high content of Ca, Fe, Cu, Mg and P was observed. During the cooking process for 4 h with successive water changes, it was observed that all these minerals presented significant differences in their concentrations throughout the experiment in the cooked sample. However, although this process decreases the mineral content in the feed, it improves the bioavailability of Ca due to the reduction in antinutrients such as oxalic acid. The oxalic acid content was removed up to undetectable values after the 4 h cooking treatment. However, the phytate was only 40% removed. The results show that *C. retusa* pods can be a source of minerals (Ca, Fe, Cu and Mg) under controlled cooking conditions and decreases in antinutrients such as oxalic acid.

Author Contributions: Conceptualization, L.M.; methodology, A.S. and O.H.; software, L.M.; validation, O.H. and R.V.; formal analysis, A.S. and F.F., investigation, A.S., resources, L.M., data curation, S.C.; writing—original draft preparation, L.M.; writing—review and editing, S.C.; visualization, L.M.; supervision, O.H.; project administration, L.M.; funding acquisition, L.M. All authors have read and agreed to the published version of the manuscript.

Funding: This research received no external funding. The APC was funded by Universidad de Lima.

Institutional Review Board Statement: Not applicable.

Informed Consent Statement: Not applicable.

Data Availability Statement: Data are contained within the article.

Acknowledgments: This work was supported by grant from Ia ValSe-Food CYTED Network and Tucos Factory E.I.R.L.

Conflicts of Interest: The authors declare no conflict of interest.

References

1. Abraham de Noir, F.; Bravo, S. *Frutos de leñosas nativas de Argentina*; 1a; Universidad Nacional de Santiago del Estero—UNSE, Facultad de Ciencias Forestales: Santiago del Estero, Argentina, 2014; ISBN 978-987-1676-19-4.

2. Schmeda-Hirschmann, G.; Theoduloz, C.; Jiménez-Aspee, F.; Echeverría, J. Bioactive constituents from south american prosopis and their use and toxicity. *Curr. Pharm. Des.* **2020**, *26*, 542–555. [CrossRef] [PubMed]
3. Arenas, P.; Scarpa, G.F. Edible Wild Plants of the Chorote Indians, Gran Chaco, Argentina. *Bot. J. Linn. Soc.* **2007**, *153*, 73–85. [CrossRef]
4. Mereles, L.; Caballero, S.; Coronel, E.; Villalba, R.; López, J.; Piris, P.; Wiszovaty, L.; Delmás, G.; Friesen, A. Recursos alimentarios autóctonos del Chaco. Una mirada a su potencial nutritivo. *Rojasiana*; Special issue No. 6. 2022, pp. 1–65. San Lorenzo, Paraguay. Available online: https://qui.una.py.vxsct57016.avnam.net/wp-content/uploads/2024/07/1.-ROJASIANA-Serie-especial-6-DIGITAL-FINAL-.pdf (accessed on 12 November 2024).
5. Sathe, S.; Venkatachalam, M. Influence of Processing Technologies on Phytate and Its Removal. In *Food Phytates*; CRC Press: Boca Raton, FL, USA, 2001; p. 32. ISBN 978-0-429-13488-3.
6. Horwitz, W.; Chichilo, P.; Reynols, H. *Official Methods of Analysis of the Association of Official Analytical Chemists*, 17th ed.; AOAC: Gaithersburg, MA, USA, 2000.
7. Dona, A.-M.; Verchère, J.-F. Analytical Applications of Oxocarbons. Part 3. Specific Spectrophotometric Determination of Oxalic Acid by Dissociation of the Zirconium(IV)–Chloranilate Complex. *Analyst* **1991**, *116*, 533–536. [CrossRef]
8. Iqbal, A.; Anwar, F.; Nadeem, R.; Sultana, B.; Mushtaq, M. Proximate Composition and Minerals Profile of Fruit and Flower of Karir (Capparis Decidua) from Different Regions of Punjab (Pakistan). *Asian J. Chem.* **2014**, *26*, 360–364. [CrossRef]
9. Pujol, A.; Sanchis, P.; Grases, F.; Masmiquel, L. Phytate Intake, Health and Disease: "Let Thy Food Be Thy Medicine and Medicine Be Thy Food". *Antioxidants* **2023**, *12*, 146. [CrossRef]
10. Abdulwaliyu, I.; Arekemase, S.O.; Adudu, J.A.; Batari, M.L.; Egbule, M.N.; Okoduwa, S.I.R. Investigation of the Medicinal Significance of Phytic Acid as an Indispensable Anti-Nutrient in Diseases. *Clin. Nutr. Exp.* **2019**, *28*, 42–61. [CrossRef]

biology and life sciences forum

MDPI

Proceeding Paper

Acceptability of Tortillas and Tamales Made with Nixtamalized Corn with Germinated Chia Flour †

María Eunice Cota * and Elsa Julieta Salazar *

Escuela de Nutrición, Facultad de Ciencias Químicas y Farmacia, Universidad de San Carlos de Guatemala, Guatemala City 01012, Guatemala

* Correspondence: eunisnutr49@gmail.com (M.E.C.); julietasalazar10@gmail.com (E.J.S.)

† Presented at the VI International Congress la ValSe-Food, Lima, Peru, 23–25 September 2024.

Abstract: Corn, *Zea mays*, is an ancestral food, culturally included in different forms in the Guatemalan diet. The most common form is in tortillas and tamales, which makes them suitable for incorporating other ingredients that increase their nutrient content. A sensory study was conducted with the aim of determining whether the appearance and texture of the tortilla and tamale remain acceptable when adding germinated chia seed flour (*Salvia hispanica* L). Germination was carried out for one day at 20 °C; it was prepared as flour and mixed with nixtamalized corn flour in a ratio of 10:90 and enough water to mold the tortillas; for the tamale, the same proportion of corn flour and germinated chia flour was used, and water and 8.6 percent oil were added. The tortillas and tamales were prepared and cooked in a traditional way by experts. For the acceptability test, 52 consumers were recruited, who signed the informed consent and subsequently evaluated the appearance and texture using a five-point hedonic scale (1 = I like it very much, 5 = I dislike it very much). The results indicate that the average acceptability of the appearance of the tortilla is 1.2 and the texture is 1.8. The average acceptability of the appearance of the tamale is 1.9 and the texture is 1.19. When comparing the acceptability of the appearance and texture of both preparations, a significant difference was found ($p < 0.05$), with the appearance and texture of the tamale being more acceptable.

Keywords: acceptability; corn; tamale; chia; tortilla

1. Introduction

The corn grain is one of the oldest known, approximately 7000 years old, and it originated from Central America and was an important food in the Mayan civilization. It has a role in religious beliefs and festivities [1].

This cereal is a pillar in the food security of Guatemalan families, especially those living in rural areas [2]. The seed can be consumed in the diet in the form of tortillas and tamales; however, corn contains a low protein content, approximately 8% [1]. On the other hand, the chia seed is consumed in the form of a whole seed as an ingredient in other preparations enriching the sensory characteristics of foods. Current research proposes that germination improves nutritional values in this seed, increasing the bioavailability of proteins by the action of proteases [3]. For this reason, it is intended to take advantage of the fact that corn tortillas and tamales are culturally consumed in Guatemala to incorporate germinated chia seeds and thus take advantage of their nutritional benefits; however, the sensory impact that these have on tortillas and tamales is unknown. The objective of this research was to determine if the appearance and texture of the corn tortilla and tamale remain acceptable when adding germinated chia seed flour.

2. Materials and Methods

2.1. Seeds and Corn Flour

Chia seeds (*Salvia hispanica* L.) and nixtamalized corn flour were used, obtained in local markets in Guatemala City.

2.2. Germination

Germination conditions were established as analyzed by Salgado et al. [4]. 450 g of seeds were used, which were disinfected with 96% ethanol; the mucilage was hydrated with 250 mL of distilled water. The seeds were then placed in aluminum trays where they were allowed to germinate for two days at 21 °C, with a light period of 12 h and a dark period of 12 h.

2.3. Preparation of Germinated Chia Flour

The trays with the sprouts were dried at 40 °C for 24 h in convection ovens (BOV V70F Biobase, Haan, Germany). The material was ground after drying in an ultracentrifugal mill (Mill ZM 200, Retsch®, Wolfenbüttel, Germany) at 600 rpm. The flour was stored in hermetic conditions in properly labeled ziploc bags.

2.4. Preparation of Germinated Chia Flour

2.4.1. Tortillas

They were prepared by experts in a traditional way; each tortilla was made with 90 g of nixtamalized flour and 10 g of germinated chia flour. The ingredients were mixed with enough water to obtain dough, which was molded into a circular and flat shape by hand. The tortillas were cooked at high temperatures (170–212 °C) for three minutes (one and a half minutes per side) on an iron griddle, special for tortillas, and bathed with a layer of water and calcium hydroxide solution.

2.4.2. Tamales

Prepared with the same proportion of corn flour and germinated chia flour used in tortillas, however, in addition to water, 8.6% sunflower oil and 1% NaCl were added to the formulation. The ingredients were mixed until a dough was obtained, wrapped in corn leaves, previously washed and dried. Each tamale was cooked in an aluminum pot, submerged in water, for 30 min at a temperature of approximately 100 °C.

2.5. Selection of Judges for Sensory Analysis

Using a non-probabilistic sampling, 52 consumers of corn tortillas and tamales, over 18 years of age, were chosen, a criterion chosen based on knowledge in perceiving variability in aspects of texture and appearance of tamales and nixtamalized corn tortillas. The voluntary participation of the panelists was recorded by informed consent.

2.6. Sensory Analysis

The test was carried out in a single session. Each panelist was given a tortilla and a tamale with germinated chia flour on a transparent plate. Each judge was told to remain silent during the exercise and to taste the preparations once. They were then evaluated using a data collection form and a 5-category hedonic scale with ordinal responses. According to Ruiz Vásquez and Soriano Colchado [4], hedonic scales with different numbers of categories, from "I like it a lot", "I neither like it nor dislike it", to "I dislike it a lot", are ideal for evaluating acceptability.

2.7. Statistical Analysis

The data were tabulated and processed in STATGRAPHICS Centurion XVIII. The descriptive categories on the hedonic scale were transformed into numerical scores such that 1 represents "I like it very much", 2 "I like it", 3 "I neither like it nor dislike it", 4 "I dislike it", and 5 "I dislike it very much". The numerical scores for each attribute were analyzed using an analysis of variance (ANOVA) and its respective Fisher's least significant difference (LSD) test. The criterion for the interpretation of means for texture and appearance was 0.0–2.0 acceptance, 2.1–2.9 interference, and 3.0–5.0 rejection.

3. Results

Table 1 shows the analysis of variance with its corresponding Fisher LSD test, indicating that the acceptability in appearance and texture of the tortilla with a formulation of 90 g of flour and 10 g of germinated flour, measured by hedonic scale, is acceptable, presenting a mean of 1.92 (0.82 SD) for appearance and 1.79 (0.68 SD) for texture. As for the tamale made with the same proportion of corn flour and chia sprout flour, its acceptability in appearance and texture was also considered acceptable, presenting means of 1.58 (0.60 SD) and 1.19 (0.16 SD), respectively. Rejection was not present for either preparation; however, the tortilla was the one that obtained the lowest score in acceptability for both attributes.

Table 1. Hedonic scale acceptability analysis for appearance and texture of tortillas and tamales with incorporated chia flour according to analysis of variance and their respective Fisher's LSD test.

	Appearance Attribute	
Product	**Tortilla**	**Tamale**
Formulation	90:10	90:10
Mean ± SD	1.92 ± 0.82 [a]	1.58 ± 0.60 [b]
Sum of squares	3.11	
Gl	1	
Mean square	3.12	
F-ratio	8.88	
p	0.0044	
	Texture Attribute	
Product	**Tortilla**	**Tamale**
Formulation	90:10	90:10
Mean (SD)	1.79 ± 0.68 [a]	1.19 ± 0.16 [b]
Sum of squares	9.24038	
Gl	1	
Mean square	9.21	
F-ratio	30.88	
p	1.00×10^{-6}	

Note: The formulation being 90:10 it would be 90% nixtamalized corn flour, 10% germinated chia flour. SD: standard deviation, Gl: degrees of freedom. Fisher's significant difference (LSD) test indicates that there is a significant difference between the appearance and texture of the tamale and tortilla, $p < 0.05$. [a,b] The data with different literals indicate that there is a significant difference between them according to Fisher's LSD test. Criteria: 0.0–2.0 acceptance, 2.1–2.9 interference and 3.0–5.0 rejection.

The test determined that there was a significant difference ($p < 0.05$) between the evaluated appearance of the tortilla and the evaluated appearance of the tamale, with a degree of freedom of 1 and $p = 0.0044$, with the appearance of the latter being more acceptable. It was also determined, with 1 degree of freedom and $p = 1.00 \times 10^{-6}$, that for the tortilla and tamale pair, there was a significant difference in texture, with the tamale being more acceptable in this aspect than the tortilla.

Figure 1 shows in general terms the scores of the hedonic scale of acceptability for appearance, noting that the highest score in the two attributes was obtained by the tamale with 10 g of chia flour, according to the Tukey test. The same behavior is observed in Figure 2, where the highest score on the hedonic scale for the texture attribute was for the tamale.

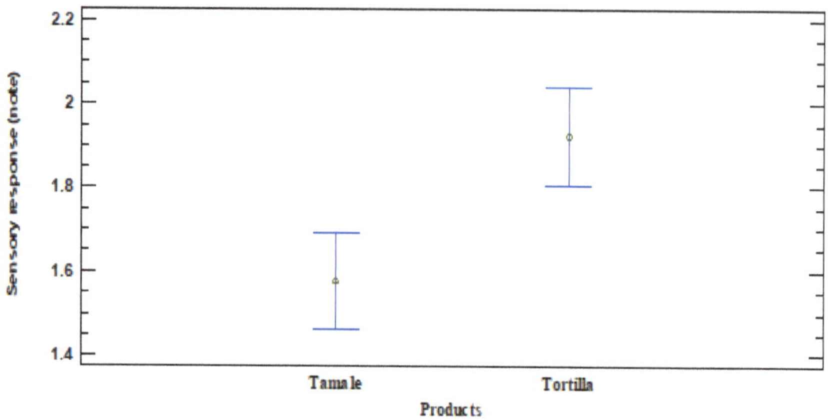

Figure 1. Hedonic scale acceptability analysis of tamales and tortillas with chia seed aggregate for the attribute appearance, according to Tukey test.

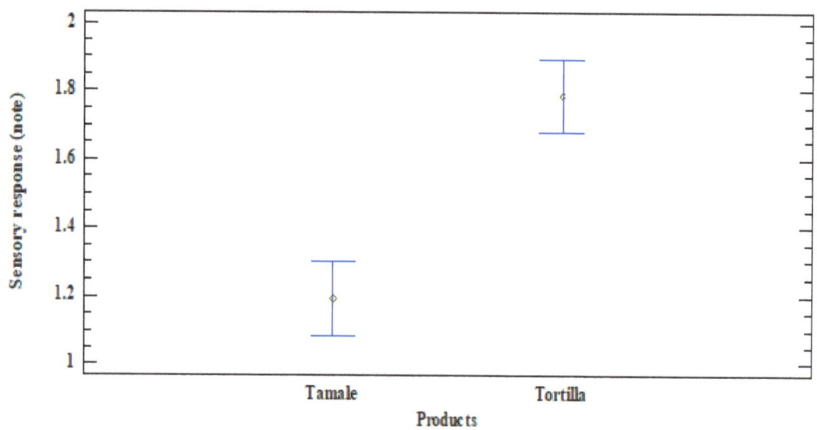

Figure 2. Hedonic scale acceptability analysis of tamales and tortillas with added chia seeds for the texture attribute, according to the Tukey test.

4. Discussion

Salgado et al. [5] state that the scientific field still does not have extensive knowledge about the incorporation of sprouted chia in food products and that it has only been sensorially evaluated by Argüelles-López et al. [6] in two functional beverages, the first using a mixture of 70% extruded amaranth flour and 30% sprouted chia flour and the second with 70% sprouted amaranth flour and 30% sprouted chia flour. They obtained as a result a high acceptability (average of 83–85), corresponding to *I like it a lot* for both beverages. With this, it could be said then that sprouted chia seed improves in general terms the acceptability of products.

On the other hand, if we talk only about chia flour, the literature reports its use to improve the sensory properties of cereals such as wheat bread, cookies, and other flours [7]. In a study carried out by Rendon-Villalobos et al. [8], where they sensorially evaluated aspects of general acceptability, flavor, color, and aroma intensity of corn tortillas made with oil and supplemented with 5%, 10%, 15%, and 20% chia seed flour, they found that this did not alter the sensory properties of the tortillas.

Other authors found higher scores in general acceptability when tortillas made with sprouted whole wheat flour were made, concluding that these provide a better appearance and greater acceptability by the consumer [9]. The conclusions of these investigations can explain why tortillas and tamales with sprouted chia flour were acceptable in the general appearance of the product, as indicated in Table 1.

As for the acceptable result in texture of the tortillas and tamales made with sprouted chia seeds in the study, it can be compared with that obtained by Ghafoor et al. [10] since they obtained a high acceptability score, in the texture attribute, when 2.0% and 3.5% of chia flour are used in bread formulations with 75.0 or 80.5% quinoa flour.

The good score for texture is also demonstrated by other authors, where they mention that replacing rice and soy flours with 2.5% whole grain chia flour does not show significant differences in acceptability of texture and appearance compared to its control, that is, it does not alter the texture of the standard product [11].

The reason why the tamale with chia seed sprout flour is more preferred in appearance and texture than the tortilla with the same formulation is due to the use of oil; this is explained by Valenzuela et al. [12]. These authors mention that fat and oils, from a technological point of view, provide palatability and flavor to foods, to the point of giving a feeling of fullness. Rodríguez-Huezo et al. [3] conclude that hydrogenated vegetable shortening improves the textural and viscoelastic properties of tamales.

5. Conclusions

It is concluded that the tortilla and tamale made from nixtamalized corn with chia sprout flour were accepted in the attributes of appearance and texture, with the tamale being preferred over the tortilla in the sensory characteristics evaluated.

Author Contributions: Conceptualization, M.E.C. and E.J.S.; methodology, M.E.C. and E.J.S.; software, M.E.C.; validation, M.E.C. and E.J.S.; formal analysis, M.E.C.; investigation, M.E.C.; resources, M.E.C. and E.J.S.; data curation, M.E.C.; writing—original draft preparation, M.E.C.; writing—review and editing, M.E.C.; visualization, M.E.C.; supervision, E.J.S.; project administration, M.E.C. All authors have read and agreed to the published version of the manuscript.

Funding: This research received no external funding.

Institutional Review Board Statement: Ethical review and approval of this study was waived because commonly consumed foods from the evaluated group were used, and the authorization of each participant was obtained through an informed consent that specified details of the research.

Informed Consent Statement: Informed consent was obtained from all subjects involved in the study.

Data Availability Statement: Informed consent forms and hedonic scale forms are available in physical form, which for the sake of the integrity of the participants are only available if requested by them via email. The processed data can also be requested via email from the authors.

Acknowledgments: This work was supported by Escuela de Nutrición, Facultad de Ciencias Químicas y Farmacia, Universidad de San Carlos de Guatemala. The presentation at the VI International Conference la ValSe-Food was supported by Universidad de Lima.

Conflicts of Interest: The authors declare no conflict of interest.

References

1. *El Maíz En La Nutrición Humana*; FAO: Rome, Italy, 1993. Available online: https://www.fao.org/4/t0395s/T0395S00.htm#Contents (accessed on 17 April 2024).
2. Chávez, O.F. Situación del cultivo de maíz en Guatemala: Principales limitaciones en la productividad. *ACI Av. Cienc. Ing.* **2022**, *14*, 1–12. [CrossRef]
3. Rodríguez-Huezo, M.E.; Flores-Silva, P.C.; Garcia-Diaz, S.; Meraz, M.; Vernon-Carter, E.J.; Alvarez-Ramirez, J. Effect of fat type on starch and protein digestibility of traditional tamales. *Die Starke* **2018**, *70*, 1700286. [CrossRef]
4. Ruiz Vásquez, D.A.; Soriano Colchado, J.L. Efecto de proporción de pasta de cacao (*Theobroma Cacao* L.) y harina de plátano (musa paradisiaca aab) en la aceptabilidad general de una mezcla alimenticia. *Cientifi-k* **2014**, *2*, 33–43.

5. Salgado, V.d.S.C.N.; Zago, L.; Antunes, A.E.C.; Miyahira, R.F. Chia (*Salvia hispanica* L.) Seed Germination: A Brief Review. *Plant Foods Hum. Nutr.* **2022**, *77*, 485–494. [CrossRef] [PubMed]
6. Argüelles-López, O.D.; Reyes-Moreno, C.; Gutiérrez-Dorado, R.; Sánchez-Osuna, M.F.; López-Cervantes, J.; Cuevas-Rodríguez, E.O.; Milán-Carrillo, J.; Perales-Sánchez, J.X.K. Functional beverages elaborated from amaranth and chia flours processed by germination and extrusion. *Biotecnia* **2018**, *20*, 135–145. Available online: https://biotecnia.unison.mx/index.php/biotecnia/article/view/721/289 (accessed on 28 April 2024). [CrossRef]
7. Katunzi-Kilewela, A.; Kaale, L.D.; Kibazohi, O.; Rweyemamu, L.M. Nutritional, health benefits and usage of chia seeds (*Salvia hispanica*): A review. *Afr. J. Food Sci.* **2021**, *15*, 48–59. [CrossRef]
8. Rendón-Villalobos, R.; Ortíz-Sánchez, A.; Solorza-Feria, J.; Trujillo-Hernández, C.A. Formulation, Physicochemical, Nutritional and Sensorial Evaluation of Corn Tortillas Supplemented with Chía Seed (*Salvia hispanica L.*). *Czech J. Food Sci.* **2012**, *30*, 118. [CrossRef]
9. Liu, T.; Hou, G.G.; Cardin, M.; Marquart, L.; Dubat, A. Quality attributes of whole-wheat flour tortillas with sprouted whole-wheat flour substitution. *Lebensm. -Wiss. Und Technol. [Food Sci. Technol.]* **2017**, *77*, 1–7. [CrossRef]
10. Ghafoor, K.; Al Juhaimi, F.; Özcan, M.M.; Uslu, N.; Ahmed, I.A.M.; Babiker, E.E. The effect of boiling, germination and roasting on bioactive properties, phenolic compounds, fatty acids and minerals of chia seed (*Salvia hispanica* L.) and oils. *Int. J. Gastron. Food Sci.* **2022**, *27*, 100447. [CrossRef]
11. Rodríguez Paucar, G.N.; Aguirre Vargas, E.B.; Villanueva Lopez, E.; León Lopez, A.; Urbina Castillo, K. Incorporation of chia seeds (*Salvia hispanica L.*) in cereal flour mixtures: Rheology and quality of sliced bread. *Dyna* **2021**, *88*, 109–116. [CrossRef]
12. Valenzuela, B.A.; Sanhueza, C.J.; Nieto, K.S. ¿es posible mejorar la calidad nutricional DE Los aceites comestibles. *Rev. Chil. Nutr. Órgano Of. La Soc. Chil. Nutr. Bromatol. Toxicol.* **2002**, *29*, 174–180. [CrossRef]

biology and life sciences forum

MDPI

Proceeding Paper

Physicochemical Properties of French Fries After Several Cycles of Frying with Moringa or Olive Oil †

Tomás Sesé [1], Lourdes Cervera-Chiner [1], Francisco José García-Mares [2], María Luisa Castelló [1] and María Dolores Ortolá [1,*]

1 Food Engineering Research Institute—FoodUPV, Universitat Politècnica de València, Camino de Vera s/n, 46022 Valencia, Spain; tosegui@etsiamn.upv.es (T.S.); loucerch@upv.edu.es (L.C.-C.); mcasgo@upv.edu.es (M.L.C.)

2 Department of Hydraulic Engineering and Environment, Universitat Politècnica de València, Camino de Vera s/n, 46022 Valencia, Spain; fjgarcia@upv.edu.es

* Correspondence: mdortola@upv.edu.es

† Presented at the VI International Congress la ValSe-Food, Lima, Peru, 23–25 September 2024.

Abstract: *Moringa oleifera* oil is characterized by its high content of oleic acid (ω9), very similar to olive oil. Moreover, it is rich in linolenic acid (ω3) and behenic acid, among other fatty acids. Furthermore, this plant has lower agronomic requirements, and it resists high temperatures. Given the current geopolitical and climatic situation, several commonly consumed oils have suffered a price increase, making them less affordable to the population. Therefore, the aim of this work was to compare the properties of French fries obtained with moringa oil or olive oil after several frying cycles, in addition to assessing their sensorial acceptance. Fried potatoes were characterized in terms of mass variation, moisture, water activity (a_w), and optical and mechanical properties. The results showed that the potatoes fried with moringa oil lost less weight during frying, which was linked to the evaporation of water during the frying stage combined with the gain of oil. However, in all cases, the a_w was similar. The color was not affected by the type of oil used, but luminosity was lower after the third frying cycle in the case of potatoes fried with moringa oil. Mechanical properties were not affected by the type of oil applied. Finally, at the sensory level, the judges evaluated all samples above 5 points, penalizing the attributes of the first-cycle moringa oil fries for being too low.

Keywords: frying; *Moringa oleifera*; oil; potatoes; quality; sensory

1. Introduction

Moringa oleifera is a plant that has demonstrated remarkable adaptability to adverse climatic conditions, making it a viable option for sustainable agriculture in the context of climate change. Originally from northern India, moringa has spread to various tropical and subtropical regions of the world, including Africa, Asia, and Latin America. Its ability to grow in poor soils and withstand long periods of drought positions it as a resilient plant of great value to communities vulnerable to the effects of climate change [1]. In this context, its introduction in the Mediterranean basin, where there is a notable increase in temperatures along with a precipitation deficit, could be an alternative to traditional crops.

Moringa seeds are especially valued for their high oil content, which represents approximately 40% of their gross weight. This oil is rich in unsaturated fatty acids, mainly oleic acid (omega 9), which gives it outstanding nutritional and medicinal properties [2]. In addition to its use in food, moringa oil is used in the cosmetic and pharmaceutical industries due to its antioxidant, anti-inflammatory, and emollient properties. Recent studies have shown that moringa oil can help protect the liver, improve cardiovascular health, and combat bacterial and fungal infections [3].

The global consumption of potatoes is significant, with over a billion people regularly including them in their diet, making potatoes the third most important food crop worldwide [4]. Among the various forms of potato consumption, fried potatoes, such as French fries and potato chips, are particularly popular. However, the frying process can significantly impact the quality of the oil used. Frying at high temperatures leads to the degradation of oil, resulting in the formation of harmful compounds and a decrease in nutritional quality [5]. Studies have shown that the type of oil used and the frying conditions can influence the extent of these changes, affecting both the sensory properties of the fried product and its health implications. For instance, oils with higher stability, such as those rich in monounsaturated fats, are less prone to oxidation and can maintain better quality during frying [6].

Therefore, this work aimed to compare the physicochemical properties of French fries obtained with moringa oil and olive oil after several frying cycles, in addition to assessing their sensory acceptance.

2. Materials and Methods

2.1. Raw Materials

Potatoes (Alegria variety, origin Spain, 50/80 mm CAT I caliber) and extra virgin olive oil (Hojiblanca variety) purchased in a local supermarket were used. Moringa oil was extracted from moringa pod seeds grown in an experimental plot at the Universitat Politècnica de València (UPV) with a screw press at 100 °C (GBT26883-2011) at the University Institute of Food Engineering-FoodUPV.

2.2. Frying of the Potatoes

The potatoes were fried with a potato-oil ratio of 1:1.5, introducing them into the pan (16 cm diameter) when the oils reached a temperature of 160 °C and frying them for 4 min. The oil was stored in glass containers in the dark for use in the following cycles. This process was carried out in the same way for the three frying cycles analyzed.

2.3. Analytical Determinations

All analytical determinations on the potatoes and the oils were carried out before and after the frying process, at least in triplicate. The potatoes were weighed on an analytical balance before and after frying, and the change in mass was expressed as a function of the initial mass. The water content of the raw and fried potatoes was obtained by the AOAC gravimetric method (930.15.) in a vacuum oven (Selecta Vaciotem, 4001489). The water activity (a_w) of raw and fried potatoes was analyzed by a dew point hygrometer (Decagon Devices Inc of AquaLab TDL). For the analysis of the color of French fries, a colorimeter (Minolta, model CM-3600d) was used to obtain the CIELab coordinates by sextuplicate, taking as reference the observer 10° and illuminator D65. In addition, the color differences (ΔE) to fresh potatoes were calculated using Equation (1):

$$\Delta E^* = \sqrt{(\Delta a^*)^2 + (\Delta b^*)^2 + (\Delta L^*)^2} \quad (1)$$

A cutting test was carried out on the potatoes with a probe (A/BS cutting blade according to AIB method) at a feed rate of 0.5 mm/s using a universal press (Texturometer TA-XTplus Texture Analyse Aname, Godalming, UK) and analyzing the force up to the breaking point and the area under the curve. The measurements were carried out in sextuplicate.

2.4. Sensory Analysis

An organoleptic analysis of 4 French fries was carried out according to the oil used (M: moringa and O: olive) and considering one (C1) or three (C3) frying cycles. Thus, the samples chosen for this analysis were MC1, OC1, MC3, and OC3, which were coded with three-digit random numbers. The tasting was carried out by presenting the freshly fried potatoes to 14 judges. The questionnaire was carried out using the SensesBit platform, which included hedonic scale questions (9-point) [7,8] related to appearance, color, aroma, texture, taste, and overall acceptability. In addition, a penalty analysis was performed to find out whether an attribute above or below the JAR penalized the overall acceptability score.

2.5. Statistical Analysis

Statgraphics CenturionXIX.64 was used for statistical analysis of the results. An ANOVA analysis of variance was performed using the LSD (Least Significant Difference) test at a significance level of 95% (p-value $\leq$ 0.05) to determined statistical significant differences.

3. Results and Discussion

3.1. Physicochemical Properties

Figure 1 shows the weight variation suffered by the potatoes in each of the frying cycles, where it is observed that potatoes fried in olive oil lose significantly more weight than those fried with moringa oil for all cycles. This weight loss is linked to water evaporation during the frying stage combined with oil gain [9].

Therefore, in Figure 2, both the moisture and water activity of the potatoes after frying are shown.

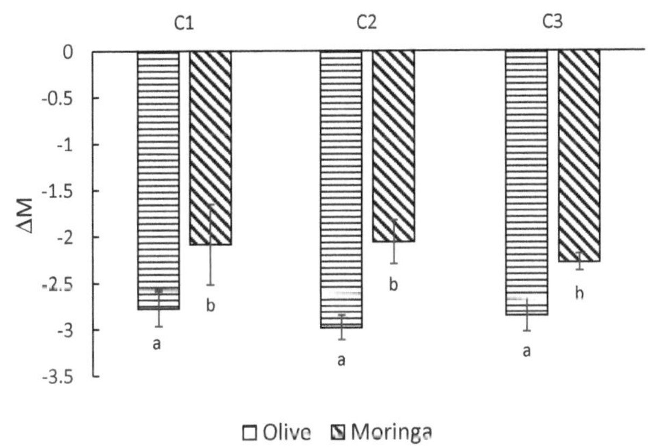

Figure 1. Potato mass variation using olive or moringa oils from different frying cycles. C1,2,3: different frying cycles. Equal letters indicate homogeneous ANOVA groups with a 95% significance level.

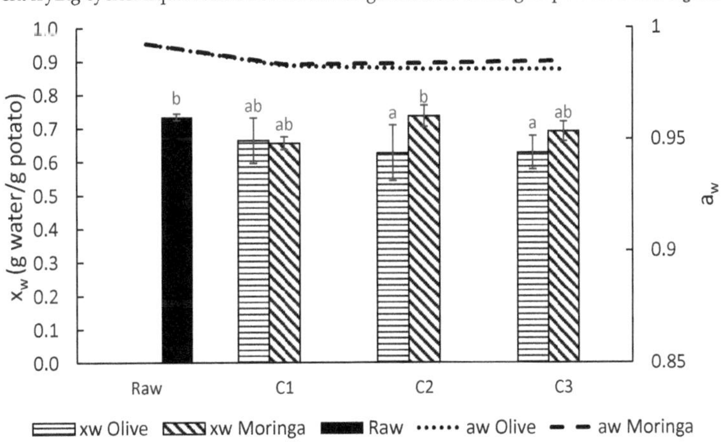

Figure 2. Variation of moisture and water activity in raw and fried potato as a function of oil and frying cycle. Equal letters indicate homogeneous ANOVA groups with a 95% significance level. C1,2,3: different frying cycles.

When moringa oil was used, no significant differences in water content were observed with respect to fresh potato in any of the frying cycles, which would indicate that weight loss would be related only to oil gain. However, in the potato fried with olive oil, there was a significant decrease in water content when oil from a second or third frying cycle was used. Therefore, and relating weight loss to the final water content of the product, frying with moringa oil would favor obtaining less fatty products.

3.2. Optical Properties

Figure 3A shows the lightness L* of the potatoes studied with respect to the raw potato as a function of the cycles.

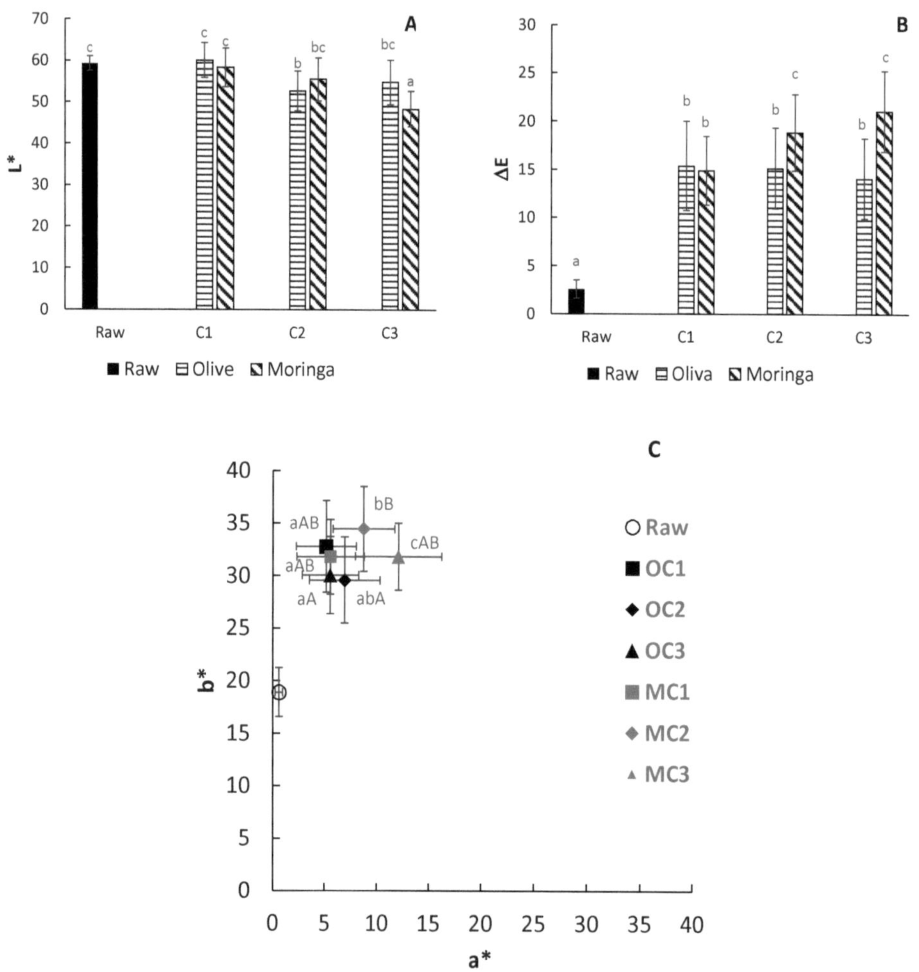

Figure 3. (**A**) Lightness (L*). (**B**) color difference (ΔE) of French fries as a function of frying cycle and oil type. (**C**) Variation of the a* and b* coordinates of the potatoes as a function of the frying cycle. Equal letters indicate homogeneous ANOVA groups with a 95% significance level, with lowercase letters for the a* coordinate and upper-case letters for the b* coordinate. C1,2,3: different frying cycles; OC1, OC2 y OC3: olive oil, frying cycle 1,2,3; MC1, MC2 y MC3: moringa oil, frying cycles 1,2,3.

As can be seen, the lightness values were similar between the potatoes fried with both oils, decreasing significantly throughout the frying cycles. As regards color variation (ΔE), in Figure 3B, the color difference is represented. Potatoes fried with both oils presented significant differences with respect to the raw potato, although these differences remained constant with frying cycles in potatoes fried with olive oil, while they increased significantly in potatoes fried in moringa oil.

On the other hand, Figure 3C shows the representation in the chromatic diagram of the a* and b* coordinates and their variation throughout the frying cycles. All values were in the first quadrant (green-yellow zone). There were no significant differences between the frying cycles of potatoes fried in olive oil, although when fried in moringa oil, the a* coordinate increased significantly with the frying cycle. The increase in the a* coordinate during frying is due to non-enzymatic browning reactions (Maillard reactions) that occur due to the presence of reducing sugars [10]. This result is in agreement with that obtained by Lalas et al. [11], who also observed an increase in the a* coordinate during several potato frying cycles.

3.3. Mechanical Properties

Table 1 shows both the maximum force applied to cut the potato and the area under the curve up to the same point.

Table 1. Maximum force and Area under the curve to the potato cut-off point.

Sample	Maximum Force (Newtons)	Area Under the Curve (N·s)
Olive C1	5.7 ± 1.9^{b}	27 ± 10^{b}
Olive C2	3.3 ± 0.5^{a}	12.2 ± 2.3^{a}
Olive C3	4.8 ± 1.9^{ab}	27 ± 9^{b}
Moringa C1	3.8 ± 1.2^{ab}	14.6 ± 6.6^{a}
Moringa C2	6.12 ± 1.5^{b}	30 ± 8^{b}
Moringa C3	5.9 ± 1.7^{b}	29 ± 11^{b}

Equal letters in the same column indicate homogeneous ANOVA groups with a 95% significance level. C1,2,3: frying cycles. OC1, OC2, OC3: olive oil, frying cycle 1,2,3; MC1, MC2, MC3: Moringa oil, frying cycles 1,2,3.

In potatoes fried with moringa oil, there was an increase in the mechanical parameters evaluated with the frying cycle, whereas non a clear tendency was observed in samples with olive oil. However, the changes recorded were very slight, so the type of oil does not seem to condition the mechanical properties of the final product. Pedreschi et al. [9] obtained similar maximum force values for French fries. According to Segnini et al. [12], the maximum breaking force is proportional to moisture content. Thus, the maximum force tends to decrease when moisture increases. Moreover, in the texture of French fries, there is great variability due to the changes in moisture and the heterogeneous distribution of starch and other compounds that affect the structure of French fries after frying [10].

3.4. Sensory Analysis

Figure 4 shows the results of the sensory analysis obtained from the survey of the attributes analyzed on a hedonic scale, both for the olive oil and moringa chips, specifically for the first and third cycles of each oil.

As can be seen, all attributes were rated higher than 5 points. On the other hand, texture was the only attribute that showed significant differences, with the potato fried with moringa oil of first frying (MC1) being the worst rated. Figure 5 shows the penalty graph of the olive oil and moringa potato chips in which the overall acceptance is related to the percentage of tasters who considered that the intensity of any of the attributes was too high or too low. In coherence with the previous results, the low intensity of all attributes was penalized for MC1 potato. In the case of MC3, only the low intensity of color and flavor was penalized. Potato fries with olive oil did not register penalties when obtained with the first frying cycle while increasing the cycles penalized flavor and texture for being too low.

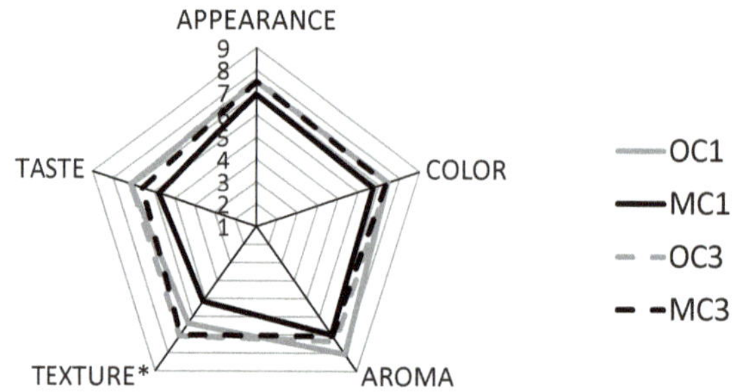

Figure 4. Results of the sensory analysis with the hedonic scale of French fries with olive oil and moringa. According to data obtained from ANOVA analysis * 95% significance level. OC1 and OC3 olive oil, frying cycle 1 and 3; MC1 and MC3: moringa oil: cycle 1 and 3.

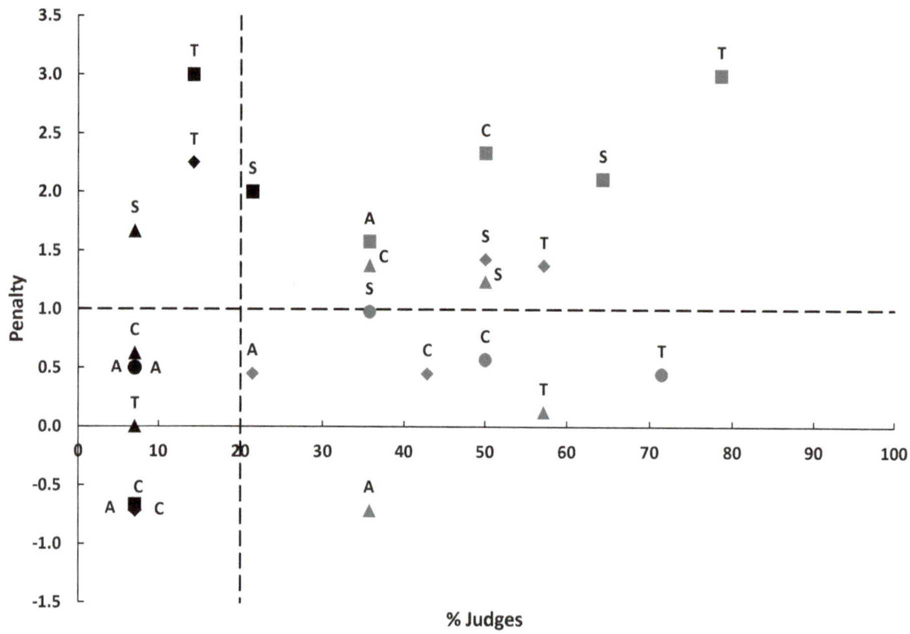

Figure 5. Graph of penalties of the fries with olive oil and moringa (Circle: OC2, Square: MC1, Rhombus: OC3 and Triangle: MC3) in the sensory analysis in the function of the analyzed attributes (T: Texture, C: Color, S: Flavor and A: Aroma). OC1 and OC3 olive oil, frying cycle 1 and 3; MC1 y MC3: moringa oil: cycle 1 and 3. Black color indicates that the intensity of the attribute is too high and the gray color is too low.

4. Conclusions

In this study, the properties of moringa oil French fries were evaluated in comparison to those fried in olive oil. During the frying process, moringa oil French fries lost less weight and water compared to olive oil-fried chips. This suggests a reduced oil uptake and, consequently, lower caloric content. Moringa oil French fries were well-received in terms of flavor and texture. Tasters highlighted their pleasant taste and crispiness, indicating that the incorporation of moringa oil did not negatively impact the sensory experience.

As frying cycles increased, moringa oil French fries developed a more intense flavor. This effect could be beneficial for products requiring a pronounced flavor profile. However, it is important to note that, compared to olive oil-fried potatoes, moringa-fried potatoes exhibited a less intense taste and aroma. This could influence consumer preference and should be considered in product design. In summary, moringa oil French fries offer a promising alternative from both a nutritional and sensory standpoint. Their lower caloric content and positive acceptance could make them an attractive option for health-conscious consumers. Nevertheless, further research is needed to fully understand their impact on final product quality.

Author Contributions: Conceptualization, M.D.O. and M.L.C.; methodology, M.D.O., M.L.C., F.J.G.-M. and L.C.-C.; formal analysis, L.C.-C. and T.S.; investigation, M.D.O., M.L.C., L.C.-C. and T.S.; resources, F.J.G.-M.; data curation, L.C.-C. and T.S.; writing—original draft preparation, L.C.-C. and T.S.; writing—review and editing, M.D.O., M.L.C. and L.C.-C.; supervision, M.D.O. and M.L.C.; project administration, M.D.O. and M.L.C.; funding acquisition, M.D.O., M.L.C., L.C.-C. and F.J.G.-M. All authors have read and agreed to the published version of the manuscript.

Funding: This work funded by the European Union-Next Generation EU by the "Margarita Salas" contract (2021–2024) of L. Cervera-Chiner from Universitat Politècnica de València and the Spanish Ministry of Universities with the UPV Contract and the APC was funded by Universidad de Lima (Perú).

Institutional Review Board Statement: A sensory test was conducted with human subjects. The study was conducted in accordance with the Declaration of Helsinki. The data obtained were treated in accordance with the Regulation (EU) 2016/679 of the European Parliament and of the Council of April 27, 2016, on the protection of natural persons with regard to the processing of personal data and on the free movement of such data, and repealing Directive 95/46/EC (General Data Protection Regulation).

Informed Consent Statement: Informed consent was obtained from all subjects involved in the study.

Data Availability Statement: The original contributions presented in this study are included in the article. Further inquiries can be directed to the corresponding author.

Acknowledgments: The authors thank the support provided by the Universitat Politècnica de València and La ValSe-Food-CYTED (119RT0567).

Conflicts of Interest: The authors declare no conflict of interest.

References

1. Anwar, F.; Latif, S.; Ashraf, M.; Gilani, A.H. *Moringa oleifera*: A Food Plant with Multiple Medicinal Uses. *Phytother. Res.* **2007**, *21*, 17–25. [CrossRef] [PubMed]
2. Fu, X.; Su, J.; Hou, L.; Zhu, P.; Hou, Y.; Zhang, K.; Li, H.; Liu, X.; Jia, C.; Xu, J. Physicochemical and Thermal Characteristics of *Moringa oleifera* Seed Oil. *Adv. Compos. Hybrid Mater.* **2021**, *4*, 685–695. [CrossRef]
3. Horn, L.; Shakela, N.; Mutorwa, M.K.; Naomab, E.; Kwaambwa, H.M. *Moringa oleifera* as a Sustainable Climate-Smart Solution to Nutrition, Disease Prevention, and Water Treatment Challenges: A Review. *J. Agric. Food Res.* **2022**, *10*, 100397. [CrossRef]
4. Singh, J.; Kaur, L. (Eds.) *Advances in Potato Chemistry and Technology*, 4th ed.; Academic Press: Cambridge, MA, USA, 2016.
5. Wang, Y.; Wu, X.; McClements, D.J.; Chen, L.; Miao, M.; Jin, Z. Effect of New Frying Technology on Starchy Food Quality. *Foods* **2021**, *10*, 1852. [CrossRef] [PubMed]
6. Santos, C.S.P.; Cruz, R.; Cunha, S.C.; Casal, S. Effect of Cooking on Olive Oil Quality Attributes. *Food Res. Int.* **2013**, *54*, 2016–2024. [CrossRef]
7. *UNE 87025:1996*; Análisis Sensorial. Metodología. Perfil de Textura. Asociacion Española de Normalización: Madrid, Spain, 1996.
8. *ISO 4121:2003*; Sensory Analysis—Guidelines for the Use of Quantitative Response Scales. International Organization for Standardization: Geneva, Switzerland, 2003.
9. Pedreschi, F.; Segnini, S.; Dejmek, P. Evaluation of the Texture of Fried Potatoes. *J. Texture Stud.* **2004**, *35*, 277–291. [CrossRef]
10. Pedreschi, F.; Moyano, P.; Kaack, K.; Granby, K. Color Changes and Acrylamide Formation in Fried Potato Slices. *Food Res. Int.* **2005**, *38*, 1–9. [CrossRef]

11. Lalas, S.; Gortzi, O.; Tsaknis, J. Frying Stability of Moringa Stenopetala Seed Oil. *Plant Foods Hum. Nutr.* **2006**, *61*, 99–108. [CrossRef] [PubMed]
12. Segnini, S.; Dejmek, P.; Öste, R. Reproducible Texture Analysis of Potato Chips. *J. Food Sci.* **1999**, *64*, 309–312. [CrossRef]

biology and life sciences forum

MDPI

Proceeding Paper

Obtaining Carotenoids and Capsaicinoids (*Capsicum chacoense*) with a Green Solvent (*Acrocomia aculeata* Almond Oil) †

Eva Coronel, Laura Correa, Malena Russo, Carlos Zaracho, Maria Caravajal, Silvia Caballero, Rocio Villalba and Laura Mereles *

Biodiversidad, Alimentos y Salud Group (BIOALSA), Dirección de Investigación, Facultad de Ciencias Químicas, Universidad Nacional de Asunción, San Lorenzo 111421, Paraguay; ecoronel@qui.una.py (E.C.); lcorrea@qui.una.py (L.C.); malerusso96@gmail.com (M.R.); matiaszzaracho111111@gmail.com (C.Z.); marcarvajal.mbc@gmail.com (M.C.); scaballero@qui.una.py (S.C.); rvillalba@qui.una.py (R.V.)

* Correspondence: lauramereles@qui.una.py

† Presented at the VI International Congress la ValSe-Food, Lima, Peru, 23–25 September 2024.

Abstract: *Capsicum chacoense* (wild red pepper) and *Acrocomia aculeata* almond (*Paraguayan coconut*) are fruits native to Paraguay which are little-used and can be sources of important bioactive compounds. The aim of this work was to evaluate the use of Paraguayan coconut kernel oil as a green solvent for the extraction of carotenoids and capsaicinoids from wild red pepper. Ultrasound-assisted extraction was performed (solvent ratio; 0.7 g/mL, amplitude 80%, for 17 min). The freeze-dried red pepper fruit, coconut oil, and coconut+red pepper oil were characterized by total carotenoids, total capsaicinoids, total phenolic compounds (TPCs), total antioxidant capacity (TAC), fatty acid (FA) profile, and color. It was possible to extract 46.7% of the carotenoids and 42.5% of the capsaicinoids present in the red pepper. However, only about 7% of TCP and TAC were maintained in the coconut+red pepper oil obtained. In the FA profile of red pepper oil, oleic acid and palmitic acid were observed as the main FAs. Conversely, in coconut oil, lauric acid and oleic acid were observed as the main components. In coconut+red pepper oil, the same main FAs were found, but in a lower percentage of lauric acid and higher percentage of oleic acid. Based on the results, coconut oil is a green solvent for the extraction of lipophilic secondary metabolites such as carotenoids and capsaicinoids. These can provide sensory characteristics such as color and flavor to coconut oil from *Capsicum chacoense*. In the oil obtained (coconut+red pepper), a significant difference in the FA profile was also seen, where the majority FA was oleic acid.

Keywords: *Acrocomia aculeata* almond oil; capsaicinoids; *Capsicum chacoense*; green solvent; Paraguayan Chaco

1. Introduction

The *Capsicum chacoense*, or "aji del monte", is a native chili pepper that grows wild, mainly in the Chaco region, Paraguay. It is commonly used as a condiment or vegetable by indigenous peoples. However, its demonstrated content of secondary metabolites such as carotenoids and capsaicinoids makes it attractive for use in products useful for industry [1]. On the other hand, *Acrocomia aculeata*, coco mbokaja, is widely distributed throughout the national territory. Its distribution is so extensive that a large proportion of coconut almonds are currently wasted [2], highlighting the need for studies that demonstrate their utility in different fields, for example, as a green solvent.

In a world increasingly seeking healthy and sustainable food, ecological solvents with biological and renewable bases are being used for the extraction, purification, and formulation of natural and food products [3]. One alternative is vegetable oils, such as those from the Paraguayan coconut almond or mbokaja. The objective of this work was to evaluate the use of coconut almond oil as a green solvent for the extraction of total carotenoids and total capsaicinoids from *Capsicum chacoense* fruit.

2. Materials and Methods

2.1. Plant Material and Use of the Green Solvent

Raw *Acromia aculeata* almond oil was provided by a private company, then degummed, neutralized, and bleached. *Capsicum chacoense* fruits were manually collected from wild plants in Filadelfia, Boquerón, then lyophilized to a moisture content of less than 10%.

Refined coconut oil was used as a green solvent for extracting carotenoids and capsaicinoids from whole *C. chacoense* fruits. For this, an ultrasonic probe Model Q55 sonicator, (QSonica, Newtown, CT, USA) was used. The probe tip was immersed halfway into the depth level of the refined coconut almond oil mixed with ground chili in the center of the Falcon tube. The solvent ratio was 0.7 g/mL, with an amplitude of 80%, for 17 min at 40 °C. The resulting product was the extract of chili fruits with refined coconut almond oil (CO+PE).

2.2. Analytical Methods

The antioxidant potential was determined in the CO, CO+PE, and the lyophilized whole chili fruits. First, an extract was made with methanol (60:40) and acetone (70:30) for the lyophilized fruits, while for the oils, 1 g of oil was dissolved in 3 mL of 80% methanol, then sonicated and centrifuged. This procedure was repeated three times and finally brought to a volume of 10 mL. The total phenolic compounds were analyzed using the Folin–Ciocalteu reagent [4], and total antioxidant capacity by ABTS radical inhibition [5]. Additionally, the concentration of total carotenoids was determined by UV-Vis spectrophotometry [6].

The total capsaicinoids in the lyophilized whole fruit were determined as described Coronel et al. [7]. For the extraction of total capsaicinoids in the oil, 1 g of oil was first dissolved in 3 mL of 80% methanol, then sonicated and centrifuged. This procedure was repeated three times and then brought to a final volume of 10 mL. The obtained extract was injected into the HPLC, maintaining the chromatographic conditions mentioned in [7].

The fatty acid profile of CO, CO+PE, and the oil of the whole *C. chacoense* fruit (PO) extracted cold with hexane in a 3:1 ratio was analyzed. First, fatty acid esterification was performed according to the official AOCS Ce 2-66 method [8] and analyzed in a GC-FID according to the official AOCS Ce 1j-07 method [8].

2.3. Statistical Analysis

The results were expressed as means ± standard deviation (SD) from three independent replicates. To compare the samples, ANOVA and Tukey's post-test were used. Values with $p \leq 0.05$ were considered as statistically significant with the assistance of Graph Pad Prism 5.0 software (GraphPad Software, Inc., San Diego, CA, USA) for the calculations.

3. Results and Discussion

Table 1 shows the results of TPCs, antioxidant capacity by ABTS radical inhibition, total carotenoids, and total capsaicinoids in CO, CO+PE, and the whole chili fruit. Table 2 shows the fatty acid profile in CO, whole fruit oil, and CO+PE, where it was found that the major fatty acid in the three samples analyzed was oleic acid.

To determine the efficiency of coconut almond oil as a green solvent, the concentration of total carotenoids and total capsaicinoids in CO, whole fruit, and CO+PE was determined. Significant differences were found in the total carotenoid content in the three samples (ANOVA, Tukey's post-test, $p > 0.05$), with the highest value in the lyophilized fruit. It was found that the green solvent extracted 46.7% of the total carotenoids present in the fruit. Higher values of carotenoids have been reported in other hexane extracts of *Capsicum annuum* paprika [9]. However, a study conducted on crushed aji del monte from the Paraguayan Chaco reported a content of up to 239 mg/kg of total carotenoids [7], less than half of that found in the lyophilized fruits of this study (515.17 ± 62.07 mg/kg) and values similar to those found in CO+PE (237.51 ± 6.07 mg/kg). The same study reported values of up to 211 mg/100 g of capsaicinoids, lower than those found in this study.

Table 1. Antioxidant potential and total capsaicinoids of CO, whole fruit of *C. chacoense*, and CO+PE.

Parameter	CO	CO+PE	Freeze-Dried Fruits	Extraction Percentage
TPCs (mg GAE/100 g)	ND	45.84 ± 0.93 [a]	643.40 ± 18.59 [b]	7.1
ABTS (mM TEAC/g)	0.23 ± 0.06 [a]	3.03 ± 0.21 [b]	39.09 ± 2.01 [c]	7.8
Total carotenoids (mg/kg)	6.88 ± 1.55 [a]	237.51 ± 6.07 [b]	515.17 ± 62.07 [c]	46.7
Total capsaicinoids (mg/100 g)	ND	51.99 ± 2.91 [a]	122.42 ± 11.16 [b]	42.5

ND: not detectable. CO: refined coconut almond oil; CO+PE: chili pepper fruit extract with refined coconut almond oil. TPCs: total phenolic compounds. Results are expressed as mean + SD. GAE: gallic acid equivalents; mM TEAC: millimole equivalents of TROLOX. Different letters in each row indicate significant differences (ANOVA, with Tuckey's post hoc test, $p > 0.05$).

Table 2. Fatty acid profile of CO, whole fruit, and CO+PE.

	Fatty Acids	CO	Whole Fruit	CO+PE
Saturated	Hexanoic (6:00)	0.24 ± 0.07 [a]	ND	0.21 ± 0.06 [a]
	Octanoic (8:00)	3.74 ± 0.43 [a]	ND	5.21 ± 1.71 [a]
	Decanoic (10:00)	3.47 ± 0.19 [a]	ND	2.78 ± 0.94 [a]
	Lauric (12:00)	34.54 ± 1.05 [a]	ND	28.24 ± 3.67 [b]
	Myristic (14:00)	7.87 ± 0.03 [a]	0.37 ± 0.01 [b]	6.78 ± 0.55 [c]
	Palmitic (C16:00)	7.44 ± 0.18 [a]	17.47 ± 0.21 [b]	7.84 ± 0.56 [a]
	Stearic (18:00)	2.52 ± 0.03 [a]	2.24 ± 0.13 [a]	2.04 ± 0.33 [a]
Total		59.82	20.08	53.13
Monounsaturated	Palmitoleic (ω-7)	1.91 ± 0.06	ND	ND
	Oleic (ω-9)	35.52 ± 1.37 [a]	68.07 ± 0.93 [b]	40.10 ± 8.53 [a]
Total		37.43	68.07	40.10
Polyunsaturated	Linoleic (ω-3)	4.69 ± 0.15 [a]	5.98 ± 0.23 [a]	6.28 ± 0.56 [a]
	Arachidonic (ω-6)	ND	3.10 ± 0.35	ND
Total		4.69	9.08	6.28

ND: not detectable. CO: refined coconut almond oil; CO+PE: chili pepper fruit extract with refined coconut almond oil. Different letters in each row indicate significant differences (ANOVA, with Tuckey's post hoc test, $p > 0.05$).

Coconut oil was able to extract 42.5% of the capsaicinoids present in the whole fruit. It was found that TPCs are poorly extracted by coconut oil, which is expected due to the hydrophilic nature of these compounds.

Furthermore, the fatty acid profile of CO, whole fruit oil, and CO+PE was analyzed. Significant differences were found between the three samples (ANOVA, Tukey's post-test, $p < 0.05$) in the myristic acid content, with the highest value for CO (7.87 ± 0.03%). Overall, it was observed that CO+PE maintains the fatty acid ratio of CO, being the major oil, with a slight decrease in lauric acid and an increase in oleic acid, matching the fatty acid ratio found in the oil of *C. chacoense* fruits. In another study conducted on *Acrocomia aculeata* almond oil cultivated in Brazil, lauric acid (45.25%) and oleic acid (23.96%) were found to be the major fatty acids [10], different from the findings of this study (34% lauric acid and 35% oleic acid). Regarding the oil profile of the red pepper fruit, it was found that the major fatty acid, at 68%, was oleic acid. However, other studies have found that the *Capsicum* genus generally has linolenic acid as the major fatty acid [9,10].

4. Conclusions

This study demonstrates that coconut almond oil is an effective green solvent for extracting bioactive compounds from *Capsicum chacoense*. The oil extracted 46.7% of total carotenoids and 42.5% of total capsaicinoids from the fruit, showing its potential for industrial applications in producing natural antioxidant-rich products. Although less efficient in extracting phenolic content, coconut almond oil maintained the fatty acid profile of the original samples. These findings highlight the viability of using coconut almond oil as a sustainable and eco-friendly solvent.

Author Contributions: Conceptualization, E.C., L.C. and L.M.; methodology, E.C. and L.C.; software, M.R. and C.Z.; validation, M.R., C.Z. and L.C.; formal analysis, E.C., M.R., C.Z., M.C. and R.V.; investigation, S.C.; resources, L.C. and L.M.; data curation, E.C. and L.C.; writing—original draft preparation, E.C.; writing—review and editing, E.C. and L.M.; visualization, L.M.; supervision, L.M.; project administration, S.C.; funding acquisition, E.C., L.C. and L.M. All authors have read and agreed to the published version of the manuscript.

Funding: This work is co-financed by the National Council of Science and Technology (CONACYT) with the support of the FEEI, project INIC01-221 "Estudio in vitro del efecto de extractos de semillas oleaginosas del Paraguay sobre las enzimas digestivas clave en el desarrollo del síndrome metabólico". And The APC was funded by Universidad de Lima—Peru. Information regarding the funder and the funding number should be provided.

Institutional Review Board Statement: Not applicable.

Informed Consent Statement: Not applicable.

Data Availability Statement: The original contributions presented in the study are included in the article, further inquiries can be directed to the corresponding author.

Acknowledgments: This work was supported by a grant provided by Ia ValSe Food-CYTED Network, Universidad de Lima—Peru, the Consejo Nacional de Ciencia y Tecnología (CONACYT) of Paraguay, and Adeline Friesen (Tucos Factory EIRL).

Conflicts of Interest: The authors declare no conflict of interest.

References

1. Coronel, E.; Mereles, L.; Caballero, S.; Alvarenga, N. Crushed Capsicum Chacoense Hunz Fruits: A Food Native Resource of Paraguay with Antioxidant and Anthelmintic Activity. *Int. J. Food Sci.* **2022**, *2022*, 1512505. [CrossRef] [PubMed]
2. Rodríguez Bonet, S.; Chaparro Aguilera, L.; Benítez Villalba, J.; Villalba, D.; Arévalos Rotela, L.; Benítez Franco, R.; Rodríguez Bonet, S.; Chaparro Aguilera, L.; Benítez Villalba, J.; Villalba, D.; et al. Caracterización fisicoquímica de la pulpa y almendra de Acrocomia aculeata. *Investig. Agrar.* **2020**, *22*, 46–52. [CrossRef]
3. Yara-Varón, E.; Li, Y.; Balcells, M.; Canela-Garayoa, R.; Fabiano-Tixier, A.-S.; Chemat, F. Vegetable Oils as Alternative Solvents for Green Oleo-Extraction, Purification and Formulation of Food and Natural Products. *Molecules* **2017**, *22*, 1474. [CrossRef] [PubMed]
4. Singleton, V.L.; Rossi, J.A. Colorimetry of Total Phenolics with Phosphomolybdic-Phosphotungstic Acid Reagents. *Am. J. Enol. Vitic.* **1965**, *16*, 144–158. [CrossRef]
5. Re, R.; Pellegrini, N.; Proteggente, A.; Pannala, A.; Yang, M.; Rice-Evans, C. Antioxidant Activity Applying an Improved ABTS Radical Cation Decolorization Assay. *Free Radic. Biol. Med.* **1999**, *26*, 1231–1237. [CrossRef] [PubMed]
6. Instituto Interamericano de Cooperación para la Agricultura; Porgrama Cooperativo para el Desarrollo y Agroindustrial del Cono Sur. *Protocolos Estandarizados para la Valorización de Frutos Nativos del PROCISUR Frente a la Creciente Demanda por Ingredientes y Aditivos Especializados (Carotenoides, Antocianinas y Polifenoles)*; Díaz, E.D., Pino, M.T., Saavedra, J., Eds.; Instituto Interamericano de Cooperación para la Agricultura (IICA): Montevideo, Uruguay, 2018; ISBN 978-92-9248-793-5.
7. Firestone, D. *Official Methods and Recommended Practices of the American Oil Chemist's Society AOCS*, 6th ed.; AOCS: Champaign, IL, USA, 2009.
8. Kostrzewa, D.; Mazurek, B.; Kostrzewa, M.; Jóźwik, E. Carotenoids and Fatty Acids Obtained from Paprika Capsicum Annuum by Supercritical Carbon Dioxide and Ethanol as Co-Extractant. *Molecules* **2023**, *28*, 5438. [CrossRef] [PubMed]
9. da Silva, L.P.R.; Rodrigues, E.L.; Hiane, P.A.; Nunes, Â.A.; Filiú, W.F.; Cavalheiro, L.F.; Nazário, C.E.D.; Asato, M.A.; Freitas, K.d.C.; Bogo, D.; et al. Bocaiuva (*Acrocomia aculeata*) Nut Oil: Composition and Metabolic Impact in an Experimental Study. *Food Sci. Technol.* **2022**, *43*, e43522. [CrossRef]
10. Jarret, R.L.; Levy, I.J.; Potter, T.L.; Cermak, S.C. Seed oil and fatty acid composition in *Capsicum* spp. *J. Food Compos. Anal.* **2013**, *30*, 102–108. [CrossRef]

biology and life sciences forum

MDPI

Proceeding Paper

Effect of Alkaline Extrusion Temperature on Rheological Properties of Andean Corn Dough †

Natalia Ester Dominguez *, María Alejandra Gimenez, Cristina Noemi Segundo, Ileana de los Angeles Gremasqui, Manuel Oscar Lobo and Norma Cristina Samman

CIITED Centro de Investigación Interdisciplinario en Tecnología y Desarrollo Social del NOA-CONICET, Facultad de Ingeniería, Universidad Nacional de Jujuy, Ítalo Palanca 10, San Salvador de Jujuy 4600, Argentina; malejandragimenez@gmail.com (M.A.G.); segundocristina@gmail.com (C.N.S.); igremasqui@ciited.unju.edu.ar (I.d.l.A.G.); mlobo958@gmail.com (M.O.L.); normasamman@gmail.com (N.C.S.)

* Correspondence: ndominguez@ciited.unju.edu.ar

† Presented at the VI International Congress la ValSe-Food, Lima, Peru, 23–25 September 2024.

Abstract: The application of alkaline extrusion in whole corn flour not only produces partial gelatinization of starch but also favors interactions between its components and releases natural hydrocolloids, modifying the rheological properties and suitability for application in gluten-free pastas or bakery products. The intensity of these modifications and therefore their rheological quality depend on the extrusion conditions. This work aimed to study the effect of alkaline extrusion temperature (70, 80 and 90 °C) at 40% feed humidity on the rheological properties of Cuzco corn flour and its dough. The increase in extrusion temperature had a significant effect ($p < 0.05$) on the degree of gelatinization of the flours (increase from 31.74 to 71.64%), which impacted their viscous properties. The degree of gelatinization, the formation of amylose–lipid–protein complexes and the soluble fiber content modified the rheological properties of the dough, decreasing the elastic modulus with increasing extrusion temperature. The most cohesive and elastic doughs were obtained at a lower temperature (70 °C), which presented greater resistance to kneading. This study will expand the use of whole Andean corn flour in gluten-free dough to obtain pastas and/or bakery products, reducing the use of food additivess.

Keywords: alkaline extrusion; andean corn; dough; dietary fiber; gluten free; hydrocolloid

1. Introduction

Nixtamalization is a widely used technology in Central America in the processing of corn to produce "tortillas" [1]. Alkaline extrusion has emerged as a sustainable alternative to traditional nixtamalization, as an ecological hydrothermal treatment that allows for the utilization of whole grains [2]. The combination of mechanical stress with heat treatment and an alkaline agent produces changes and promotes the interaction between flour components, modifying its functional properties to be incorporated in gluten-free foods [3]. The presence of the alkaline agent solubilizes components of the pericarp, generating natural hydrocolloids that influence the rheological properties of the doughs [4]. The degree of starch gelatinization and the amylose–lipid and calcium–starch interactions are susceptible to small changes in processing conditions (moisture content and temperature); therefore, maintaining optimal extrusion conditions is important [3].

The study of the rheological properties of the dough allows for the determination of its functional characteristics. Viscoelastic parameters such as the elastic (G') and viscous (G'') modulus are fundamental in the determination of the quality of corn doughs [3] and depend on the intensity of mechanical stress that affects the cohesive and elastic properties of the doughs [2,4,5].

Hydrothermal treatments applied to gluten-free flours, such as corn flour, are essential to adjust their functional properties according to culinary and processing needs. Although these treatments have been investigated to improve the technological aptitude of gluten-free flours to develop many products [1,2,4,5], there are few studies that apply alkaline extrusion to produce fresh, laminated and easy-to-cut doughs. Therefore, the present work aimed to study the effect of alkaline extrusion temperature on the rheological properties of Andean corn flour, Cuzco race and its doughs.

2. Materials and Methods

2.1. Raw Materials

White Andean corn of the Cuzco variety was used, supplied by the CAUQUEVA cooperative (Maimará, Jujuy Province, Argentina) and the INTA-IPAF NOA (Maimará). Whole-grain milling took place in a hammer mill (Polymix PX-MFC-90 D Kinematica) until a flour with a particle size $\leq$ 450 μm was obtained, which was sieved through mesh No. 40 (ASTM-E-11-61).

2.2. Alkaline Extrusion of Corn Flours

A total of 12 h before the process, 0.25 g of Ca $(OH)_2$/100 g of flour was added to each whole-meal flour sample, and then they were conditioned at 40% moisture. Each sample was mixed for 3 min and stored in a polyethylene bag in the refrigerator at 5 °C. To obtain alkaline-extruded corn flour (HMEA), a Brabender extruder (KE 19/25D, Germany) with a single screw of a 2:1 nominal compression ratio was used. The feed and extrusion rates were set at 20 rpm and 60 rpm, respectively. The extrusion was conducted at 70, 80 and 90 °C. The extrudates were formed through a nozzle with a diameter of 3 mm and were collected on a tray for subsequent drying.

2.3. Composition of Macronutrients in Processed Flours

For the determination of macronutrients, the AOAC [6] analytical methods were employed: proteins (AOAC International, 2005b) and lipids (AOAC International, 2005c). The total dietary fiber (TDF) and insoluble dietary fiber (IDF) contents were determined according to AACC 32-05 (2000), using the enzymatic–gravimetric method. The soluble dietary fiber (SDF) content was obtained by the difference between the TDF and IDF. The experiment was carried out in triplicate.

2.4. Physicochemical Properties of Processed Flours

2.4.1. Degree of Gelatinization (DG)

The colorimetric method developed by Birch and Priestley (1973) was used, based on the formation of an amylose–iodine complex.

2.4.2. Viscous Properties of Processed Flours

The viscous properties were determined using a Rapid Visco Analyser (RVA) (RVA series 4500, Perten instruments) following the methodology of Method 76.21.01, AACC, 2000 [7].

2.4.3. Subjective Water Absorption Water (SWAC)

SWAC was determinated according to Gaitan-Martinez et al. [8].

2.5. Rheological Properties of Doughs

The loss modulus (G'') and storage modulus (G') of the rehydrated masses were determined using the SWAC methodology. The moduli were measured using a rheometer (TA Instrumen 5 AR 1000), in accordance with the methodology reported by Platt-Lucero et al. [4].

Textural Properties

A Texture Profile Analysis (TPA) was conducted on dough discs (3 cm in diameter and 1 cm in height) which were subjected to a double-compression cycle using 40% deformation of the original height with an SMSP/50 probe and a 25 kg load cell. The test speed was

0.5 mm/s. Six discs were tested for each formulation, from which the parameters of elasticity and cohesiveness were obtained.

2.6. Statistical Analysis

The data obtained were statistically treated by analysis of variance, while the means were compared by using the LSD Fisher's test at a significance level of 0.05 using the statistical software INFOSTAT—Version 2017p (Facultad de Ciencias Agropecuarias, UNC, Cordoba Argentina). All experiments were performed in triplicate, and mean values ± standard deviation were reported.

3. Results and Discussion

3.1. Macronutrient Composition of Processed Flours

Table 1 presents the macronutrient content in native and treated Cuzco corn flours. The free lipid content was found to be between 2.14 and 2.39% in all treated flours, significantly lower than the lipid content in the native flour (4.08%). No significant differences were observed between processing temperatures. The decrease in lipids during extrusion can be attributed to lipid saponification with the alkaline agent and the complexes' (amylose–lipid and ternary complexes such as starch–protein–lipid) formation from gelatinization [7]. These complexes could participate in the formation of structures that improve the extensible and cohesive properties of gluten-free doughs.

Table 1. Macronutrient composition and degree of gelatinization of processed flours.

Muestras	Free Lipids (%)	Protein (%)	TDF (%)	IDF (%)	SDF (%)	DG (%)
HCEA 70-40	2.39 ± 0.14^{b}	11.11 ± 0.14^{ab}	11.23 ± 0.01^{ab}	6.12 ± 0.03^{a}	5.11 ± 0.01^{d}	31.74 ± 0.64^{a}
HCEA 80-40	2.14 ± 0.16^{a}	11.27 ± 0.14^{bc}	11.01 ± 0.50^{a}	7.70 ± 0.14^{b}	3.31 ± 0.36^{c}	36.47 ± 0.03^{b}
HCEA 90-40	2.14 ± 0.35^{ab}	11.10 ± 0.04^{ab}	11.85 ± 0.07^{c}	9.82 ± 0.02^{c}	2.03 ± 0.05^{b}	71.64 ± 0.03^{c}
HC nativa	4.08 ± 0.11^{c}	11.09 ± 0.19^{ab}	11.55 ± 0.25^{b}	10.5 ± 0.10^{d}	1.05 ± 0.14^{a}	ND

Different letters in the same column indicate significant differences. ($p < 0.05$). TDI: total fiber dietary, IDF: insoluble fiber dietary; SDF: soluble fiber dietary; DG: degree of gelatinization; ND: no determinate.

The total dietary fiber (TDF) of the processed flours did not present significant differences ($p < 0.05$) compared to the native flour. However, an increase in soluble dietary fiber (SDF) and a decrease in insoluble dietary fiber (IDF) were observed, suggesting a conversion between them due to the alkaline agent that induces the hydrolysis of hemicellulose and forms soluble gums that improve the texture of the doughs [2]. In addition, the SDF content increased significantly ($p < 0.05$) with decreasing extrusion temperature, evidencing a range from 1.05 to 5.11, which is consistent with that reported by Tabligbohmany et al. [8], who indicated that fiber solubilization is more linked to mechanical stress than to thermal energy because the greater frictional force causes the breakdown of the chemical bonds of the macromolecules of insoluble fiber.

3.2. Physicochemical Properties of Alkaline-Extruded Flours

3.2.1. Degree of Gelatinization

The degree of gelatinization (DG) (Table 1) increased significantly ($p < 0.05$) with the processing temperature and its values ranged from 31.74 to 71.64%. Topete-Betancour et al. [9] mentioned a value of approximately 30% as adequate to obtain easy-to-laminate doughs.

3.2.2. Viscous Properties

Significant differences ($p < 0.05$) were observed in the parameters of the profile between processed and native flours (Figure 1); the last one presented higher values of VP, BD, SB, MV and FV, but lower PT values. This could be due to the fact that most of its starch granules were intact. It is observed that VP decreases ($p < 0.05$) with extrusion temperature. The lower processing temperature provides flours with a lower degree of gelatinization; therefore, they have more starch available to develop higher viscosities. The FVs were also influenced by the extrusion temperature, observing that flour HCEA 70-40 developed higher FVs than HCEA 90-40. This behavior would indicate that flours with a higher degree of gelatinization would produce very soft doughs that could fall apart.

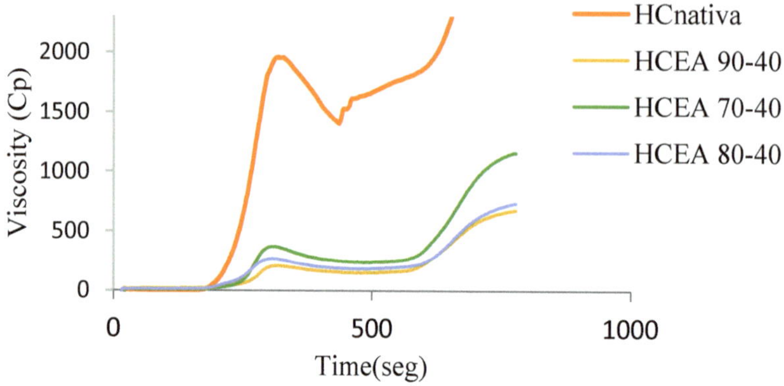

Figure 1. Viscosity profile of alkaline-extruded flours under different conditions.

3.3. Rheological Properties

In all samples, G′ > G″ was observed throughout the frequency range studied (Figure 2). This reveals that the system has a viscoelastic solid behavior, with strains that are essentially elastic and recoverable [9]. Although the dough is gluten-free, this behavior demonstrates the presence of a network with interactions that stabilize the system at the stresses utilized. The alkaline treatment causes ionization of some hydroxyl groups in the starch, allowing for the formation of Ca–starch or Ca–protein cross-linking, resulting in a stronger gel network with a higher G′ and G″ modulus. These results are in agreement with those reported by Rolandelli et al. [10].

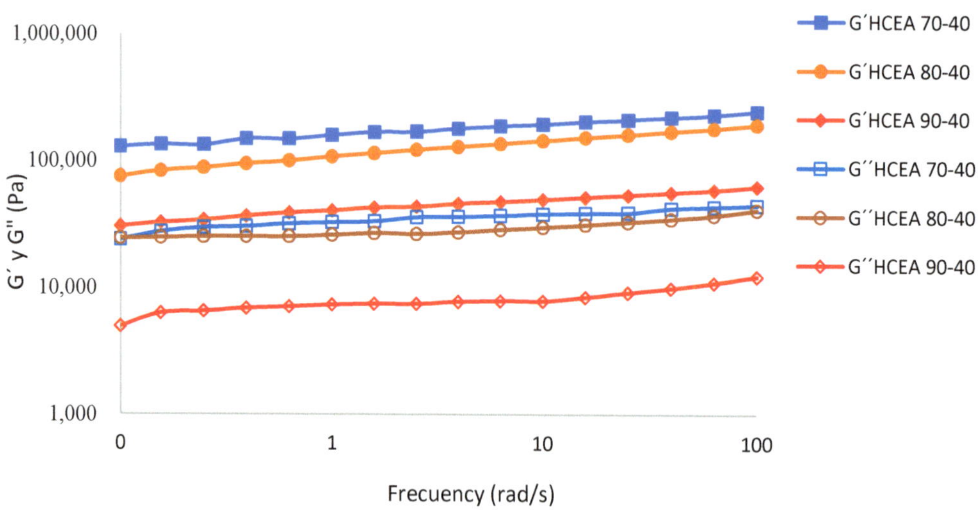

Figure 2. Rheological properties of alkaline-extruded flour mixtures obtained under different temperature conditions.

The variation in extrusion temperature changes the values of the modulus, but not the general shape. HCEA 70-40 had the highest elastic modulus possibly due to the higher soluble dietary fiber (SDF) content, improving the elasticity of the dough.

Textural Properties

Table 2 shows that SWAC values decreased with the increasing processing temperature. This behavior may be responsible for the lower elasticity and cohesiveness observed in dough elaborated with flour extruded under 90-40 conditions. The 70-40 and 80-40 flours did not present significant differences in terms of elasticity, but the second flour presented higher cohesiveness. This could be related to the higher soluble fiber content, which provides greater elasticity. The results indicate that HCEA 70-40 and HCEA 80-40 flours form more cohesive doughs. The elasticity determined in HCEA 70-40 and 80-40 (0.29 mm) was within the ranges obtained by Topete-Betancourt et al. [9].

Table 2. Subjective water absorption capacity and textural properties of processed flour doughs.

Muestras	SWAC (%)	Elasticity (mm)	Cohesive
HCEA 70-40	0.79 ± 0.00 [b]	0.29 ± 0.00 [b]	0.33 ± 0.02 [b]
HCEA 80-40	0.80 ± 0.00 [c]	0.29 ± 0.02 [b]	0.38 ± 0.04 [c]
HCEA 90-40	0.67 ± 0.01[a]	0.26 ± 0.01 [a]	0.28 ± 0.02 [a]

Different letters in the same column indicate significant differences ($p < 0.05$).

4. Conclusions

The extrusion conditions of 70 and 80 °C with 40% moisture provided flours with properties suitable for the formation of laminate and resistant doughs. These conditions favored the hydrolysis of pericarp components, generating compounds that act as hydrocolloids that interact with water, contributing to a more cohesive and elastic dough. These findings underline the relevance of carefully determining the effects of processing on the physicochemical properties of flours to ensure the quality of the final products.

Author Contributions: Conceptualization, N.E.D. and M.A.G.; methodology, N.E.D., M.A.G. and C.N.S.; software, N.E.D. and I.d.l.A.G.; validation, M.O.L. and M.A.G.; formal analysis, N.E.D. and M.A.G.; investigación N.E.D. and M.A.G.; resources, M.O.L. and N.C.S.; data curation, N.E.D., I.d.l.A.G. and C.N.S.; writing-original draft preparation, N.E.D.; writing—review and editing, M.A.G. visualization, N.E.D. and I.d.l.A.G.; supervision, M.O.L.; project administration, N.C.S.; funding acquisition, M.O.L. and N.C.S. All authors have read and agreed to the published version of the manuscript.

Funding: This research received no external funding.

Institutional Review Board Statement: Not applicable.

Informed Consent Statement: Not applicable.

Data Availability Statement: Data are conteined within the article.

Acknowledgments: This work was supported by SECTER Universidad Nacional de Jujuy—CONICET. We thank Red Ia ValSe Food-CYTED (Ref. 119RT0567) and the Universidad de Lima, Perú.

Conflicts of Interest: The authors declare no conflict of interest.

References

1. Escalante-Aburto, A.; Mariscal-Moreno, R.M.; Santiago-Ramos, D.; Ponce-García, N. An update of different nixtamalization technologies, and its effects on chemical composition and nutritional value of corn tortillas. *Food Rev. Int.* **2020**, *36*, 456–498. [CrossRef]
2. Contreras-Jiménez, B.; Gaytán-Martínez, M.; Morales-Sánchez, E.; Figueroa-Cardenas, J. Effects of tempering time, Ca $(OH)_2$ concentration, and particle size on the rheological properties of extruded corn flour. *Cereal Chem.* **2017**, *94*, 230–236. [CrossRef]
3. Enríquez-Castro, C.M.; Contreras-Jiménez, B.; Morales-Sánchez, E. Innovation in Nixtamalization by Extrusion Using the Wet Process. 2024. In *Expploring the World of Cereal Crops*; IntechOpen: Rijeka, Croatia, 2024. [CrossRef]
4. Platt-Lucero, L.C.; Ramírez-Wong, B.; Torres-Chávez, P.I.; López-Cervantes, J. Improving textural characteristics of tortillas by adding gums during extrusion to obtain nixtamalized corn flour. *J. Texture Stud.* **2010**, *41*, 736–755. [CrossRef]
5. Mohamed, I.O. Interaction of starch with some food macromolecules during the extrusion process and its effect on modulating physicochemical and digestible properties. A review. *Carbohydr. Polym. Technol. Appl.* **2023**, *5*, 100294. [CrossRef]

6. AOAC. Association of Official Analytical Chemists. International. Official Methods of Analysis. Available online: www.aoac.org (accessed on 20 January 2020).
7. *AACC International Methods*; American Association of Cereal Chemists (AACC) International: St. Paul, MN, USA, 2000.
8. Tabibloghmany, F.S.; Mazaheri Tehrani, M.; Koocheki, A. Optimization of the extrusion process through response surface methodology for improvement in functional and nutritional properties of soybean hull. *J. Food Sci. Technol.* **2020**, *57*, 4054–4064. [CrossRef] [PubMed]
9. Topete-Betancourt, A.; Santiago-Ramos, D.; Figueroa-Cárdenas, J. Relaxation tests and textural properties of nixtamalized corn masa and their relationships with tortilla texture. *Food Biosci.* **2020**, *33*, 100500. [CrossRef]
10. Rolandelli, G.; Shan, S.; Campanella, O.H. Effects of alkalinization and addition of pea protein as a co-protein to zein for the development of gluten-free doughs. *Food Hydrocoll.* **2024**, *146*, 10931. [CrossRef]

biology and life sciences forum

MDPI

Proceeding Paper

Development of Cookies Enriched with Quinoa (*Chenopodium quinoa*) and Native Collagen from Pota (*Dosidicus gigas*) Nape †

Nancy Chasquibol *, Axel Sotelo, Mateo Tapia and Rafael Alarcón

Ingeniería Industrial, Grupo de Investigación en Alimentos Funcionales, Instituto de Investigación Científica, Universidad de Lima, Av. Javier Prado Este, 4600 Fundo Monterrico Chico, Lima 33, Peru; aasotelo@ulima.edu.pe (A.S.); mtapiac@ulima.edu.pe (M.T.); ralarcor@ulima.edu.pe (R.A.)

* Correspondence: nchasquibol@ulima.edu.pe

† Presented at the VI International Congress la ValSe-Food, Lima, Peru, 23–25 September 2024.

Abstract: The giant squid (*Dosidicus gigas*) is a marine product from the Pacific Ocean. Its by-products can be used to obtain bioactive products such as collagen, proteins, and others. This work aimed to develop cookies enriched with Quinoa (*Chenopodium quinoa*) and native collagen from pota nape high in protein content, minerals, and antioxidants. Four formulations (4, 8, 12 and 16% collagen) were developed and compared with the control sample. The results showed higher protein (11.7 ± 0.3–20.8 ± 0.4%) content, lower moisture (4.7 ± 0.1–5.6 ± 0.2%), higher ash (3.0 ± 0.1–3.83 ± 0.09%), lower fat (15.29 ± 0.05–15.8 ± 0.1), and lower carbohydrate (53.89 ± 1.05–65.39 ± 0.82%) content than the control sample. Also, the cookies showed a significant content of polyphenols (618 ± 24–934 ± 23 μg gallic acid equivalent (GAE)/g), antioxidant activity (8182 ± 59–8369 ± 73 μg trolox/g) and in vitro digestibility (70.8 ± 0.1–73.6 ± 0.5%) than the control sample. The cookies also had a high mineral content: calcium (3893 ± 19 mg/kg), potassium (3222 ± 16 mg/kg), and magnesium (2108 ± 11 mg/kg). In addition, the cookies presented an adequate balance of amino acids, principally of aspartic acid, glutamic acid, serine, glycine, threonine, arginine, alanine, proline, valine, phenylalanine, and leucine. The cookies complied with the Peruvian legislation of the Healthy Law about the promotion of healthy eating for children and adolescents and with the microbiological requirements. Finally, the cookies showed a sensory acceptance of 77.8% and a shelf life of 184 days determined by the Rancimat method. The native collagen from pota nape could be used with quinoa flour to develop functional foods to help reduce child malnutrition.

Keywords: by-products; cookies; collagen; functional foods; pota

1. Introduction

Cookies are considered snacks and are popular among consumers because of their size, shape, high digestibility, high energy value, relatively low production costs, and extended shelf life [1]. At present, wheat flour is being replaced by Andean pseudocereals because of its higher protein content, improving the nutritional quality of processed products [2].

The giant squid fishery is one of the most important in Peru, with a significant increase of 46.1%. However, only 50–60% of this marine resource is harvested; its by-products could be used to obtain high-value-added bioactive products such as gelatin, collagen, chitin, protein concentrates, and essential fatty acids, among others [3]. Collagen is a protein source with functional peptides with biological activity and important health benefits such as the recovery of cartilage tissues, tendons, ligaments, and reduction in joint pain. It is also used for its techno-functional properties in food development due to its high-water absorption capacity, ability to texturize, thicken and form gels [3].

Quinoa (*Chenopodium quinoa*) is a grain native to South America, with more than 250 species whose world production is led by Peru, Bolivia and Ecuador [2]. Seven-seed flour is a mixture of oat (*Avena sativa* L.), corn (*Zea mays*), fava bean (*Vicia faba*), pea (*Pisum sativum*), cocoa (*Theobroma cacao*), quinoa (*Chenopodium quinoa*), and maca (*Lepidium meyenii*) flours, which is high in protein, amino acid and essential fatty acid, vitamin and mineral content [4]. Sacha inchi oil (*Plukenetia huayllabambana*) contains polyunsaturated fatty acids (58%) with a high content of linolenic acid (55%). Its consumption is of great importance for health to prevent cardiovascular diseases, reduce low-density lipoproteins (LDL) or cholesterol and triglycerides [5].

The objective of this research was to develop cookies with native collagen from pota nape to evaluate the effect of collagen on the proximal composition, total polyphenols, antioxidant activity (DPPH), amino acid analysis, in vitro digestibility, sensory evaluation, microbiological analysis, health law, shelf life, and mineral content.

2. Materials and Methods

2.1. Raw Materials

Native collagen from pota (*Dosidicus gigas*) nape was extracted with NaOH 0.15 N for 30 h, then dried by infrared (IRC D18, Irconfort, Seville, Spain) at 60 °C, ground (Grindomix GM200, Restch, Haan, Germany) and stored in aluminized bags until further use [6]. The ingredients used were native collagen from pota nap, flour from seven seeds (oats, corn, beans, broad beans, peas, cacao, quinoa, and maca *Lepidium meyenii*), quinoa flour, cornstarch, panela, baking powder, salt, egg, vanilla essence, sacha inchi (*Plukenetia huayllabambana*) oil, and corn oil. All ingredients were purchased at the market in Lima, Peru. The sacha inchi seeds were collected in the province of Rodríguez de Mendoza, Amazon region, Peru. Sacha inchi oil was cold-pressed at the Functional Food Laboratory of the University of Lima, Peru, and stored at 4 °C in a dark flask.

2.2. Amino Acid Profile

The amino acid profile was estimated according to Chasquibol et al. [7]. Amino acids were determined in the acid hydrolysate by high-performance liquid chromatography (Acquity Arc, Waters, Milford, MA, USA) using D, L-α-aminobutyric acid as internal standard and a 300 mm × 3.9 mm reversed-phase column (Nova Pack C18 4 μm; Waters, Milford, MA, USA), and the resultant peaks were analyzed with Empower 3 software (Waters, Santa Clara, CA, USA). The calibration curves for each amino acid were developed using a mix of the amino acid standards at the same hydrolysis conditions of the samples (Merck, Madrid, Spain). Furthermore, tryptophan content was assessed according to the method described by Yust et al. [8]. Analyzes were performed in triplicate and presented as mean values.

2.3. Cookies Formulations

The cookies were formulated according to the essential amino acid composition of seven seeds (oats, corn, beans, lima beans, peas, cocoa, quinoa, and maca), seed flour, quinoa flour, and squid nape collagen (Table 1). The powdered ingredients, including collagen, were mixed with the egg and vanilla essence for 10 min in the blender (FPSTSMPL1-053, Oster, Guandong, China). Then, corn and sacha inchi (*Plukenetia huayllabambana*) oil were added and mixed for 5 min. The cookie dough was divided into portions of 12.0 ± 0.5 g and shaped into circular cookies approximately 6 mm thick and 5 cm in diameter. The samples were baked for 13 min at 155 °C in a convection oven (HEB60R, Imaco, Guandong, China), cooled and stored in glass jars until further use.

2.4. Cookie Characterization

2.4.1. Proximal Composition

The proximal composition was carried out according to Chasquibol et al. [7]. The moisture content was determined at 110 °C to constant weight. The total protein content was determined as % nitrogen X 6.25 using a Kjeldahl analyzer (UDK 139, VELP, Usmate Velate, Italy). The ash content was determined by incineration at 550 °C for 72 h, in a muffle furnace. The fat content was determined with hexane for 4 h. Analyzes were performed in triplicate and presented as mean values.

Table 1. Cookies formulations with native collagen protein from pota (*Dosidicus gigas*) nape.

Ingredients	C	F1	F2	F3	F4
Seven-seed flour (g)	4	4	4	4	4
Quinoa flour (g)	17.6	15.6	13.6	11.6	9.6
Cornstarch (g)	6.5	6.5	6.5	6.5	6.5
Panela (g)	8	8	8	8	8
Baking Powder (g)	0.4	0.4	0.4	0.4	0.4
Salt (g)	0.4	0.4	0.4	0.4	0.4
Eggs (g)	6.5	6.5	6.5	6.5	6.5
Vainilla Essence (g)	0.6	0.6	0.6	0.6	0.6
Corn Oil (g)	3.2	3.2	3.2	3.2	3.2
Sacha Inchi oil (g)	2.8	2.8	2.8	2.8	2.8
Native collagen from pota nape (g)	0	2	4	6	8

C: Control, F1: Formulation 1, F2: Formulation 2, F3: Formulation 3, F4: Formulation 4.

2.4.2. Total Phenolic Content

The total phenolic content (TPC) was determined by Folin–Ciocalteau method [7] at 760 nm using a spectrophotometer (1280 UV–Vis Spectrophotometer Shimadzu, Kyoto, Japan). The results were expressed as µg of gallic acid equivalent (GAE)/g cookie. A calibration curve of 0.8–18 mg GAE/L was carried out. Analyzes were performed in triplicate and presented as mean values.

2.4.3. Antioxidant Activity

The antioxidant activity was determined by the DPPH method [7] with some modifications at 517 nm by spectrometry (1280 UV–Vis Spectrophotometer Shimadzu, Kyoto, Japan). The results were expressed as µg Trolox/g cookie. A calibration curve of 1.5–100 mg Trolox/L was carried out. Analyzes were performed in triplicate and presented as mean values.

2.4.4. Amino Acid Profile

The amino acid profile was determined according to Item 2.2.

2.4.5. In Vitro Digestibility (IVPD)

In vitro protein digestibility was measured according to Tapia et al. [9].

2.4.6. Microbiological Analysis

The cookies were analyzed microbiologically to stablish the microbiological criteria of sanitary quality and safety for food and beverages for human consumption [7].

2.4.7. Determination of Na, Total Sugar, Saturated Fat, and Trans-Fat

The concentrations of Na, total sugar, saturated fat, and *trans* fat were determined by official methods. Analyzes were performed in triplicate and presented as mean values.

2.4.8. Determination of Mineral Content

The concentrations of Ca, Cu, Fe, K, Mg, Na, P, Zn were determined by atomic absorption spectrophotometry (Nexion 350x, Perkin Elmer, MA, USA) according to Chasquibol et al. [7]. The results were expressed in mg/kg. Analyzes were performed in triplicate and presented as mean values.

2.4.9. Lifetime Analysis

The Rancimat method was determined at temperatures of 100, 110, and 120 °C [5].

2.4.10. Sensory Evaluation

The sensory evaluation test was performed with 30 panelists using a 9-point hedonic scale from 1 (I dislike it very much) to 9 (I like it very much) with a mean value of 5 (I neither like it nor dislike it) [7].

2.4.11. Statistical Analysis

Results were expressed as mean ± standard deviation. All measurements were determined in duplicate or triplicate. Analysis of variance (ANOVA) was used to analyze data acquired at a 95% significance level with Minitab 19.0 Software (Minitab Inc., State College, Palo Alto, CA, USA).

3. Results and Discussion

Table 2 shows the proximal composition of the cookies. The moisture (4.7 ± 0.1–5.6 ± 0.2%) and ash content (3.0 ± 0.1–3.83 ± 0.09%) of the four formulations did not present statistically significant differences to the control sample (4.43 ± 0.08%), (2.96 ± 0.01%). The fat (15.29 ± 0.05–15.8 ± 0.1%) content did not present statistically significant differences to the control sample (14.60 ± 0.02%) besides the contribution of sacha inchi (*Plukenetia huayllabambana*) oil [5].

Table 2. Proximal composition, total polyphenol, antioxidant activity, and in vitro digestibility of cookies with native collagen protein from pota (*Dosidicus gigas*) nape.

Proximal Composition	C	F1	F2	F3	F4
Moisture (%)	4.43 ± 0.08 [d]	4.70 ± 0.14 [c]	4.90 ± 0.13 [bc]	5.05 ± 0.06 [b]	5.63 ± 0.15 [a]
Ash (%)	2.96 ± 0.01 [c]	2.96 ± 0.14 [c]	3.09 ± 0.00 [c]	3.30 ± 0.05 [b]	3.83 ± 0.09 [a]
Fat (%)	14.60 ± 0.02 [c]	15.29 ± 0.05 [b]	15.38 ± 0.08 [b]	15.78 ± 0.03 [a]	15.83 ± 0.10 [a]
Protein (%)	8.33 ± 0.00 [e]	11.66 ± 0.25 [d]	15.77 ± 0.14 [c]	18.57 ± 0.44 [b]	20.82 ± 0.40 [a]
Carbohydrates (%)	69.69 ± 0.66 [a]	65.39 ± 0.82 [b]	60.86 ± 0.49 [c]	57.30 ± 0.82 [d]	53.89 ± 1.05 [e]
Total polyphenol (μg GAE/ g)	1174.21 ± 53.78 [a]	934.52 ± 23 [b]	956.18 ± 48.23 [b]	873.11 ± 78.08 [b]	618.61 ± 24.06 [c]
Antioxidant Activity DPPH (μg trolox/g)	8525.97 ± 207.98 [a]	8181.79 ± 59.49 [a]	8323.34 ± 109.55 [a]	8420.88 ± 200.11 [a]	8368.92 ± 73.02 [a]
In vitro digestibility (%)	72.54 ± 0.51 [ab]	70.82 ± 0.13 [b]	74.62 ± 0.90 [a]	72.99 ± 0.13 [a]	73.62 ± 0.51 [a]

Results are expressed as means ± SD (n = 3). [a,b,c,d,e] values in the same row with different letters differing significantly when $p < 0.05$.

The protein (11.7 ± 0.3–20.8 ± 0.4%) content in the cookie formulations was statistically higher than the control sample (8.3 ± 0.0%) due to the contribution of native collagen protein from the pota (*Dosidicus gigas*) nape. Formulation F4 was selected because of its high protein (20.82 ± 0.40%) content, and according to Codex Alimentarius CAC/GL 23-1997 it could be declared as a protein-rich food for children under 5 years of age.

The total polyphenol content of formulations with collagen, (618 ± 24 μg GAE/g to 956 ± 48 μg GAE/g) was statistically lower than the control sample (1174 ± 53 μg GAE/g). This is due to high content of quinoa in the control sample, which contains a high amount of polyphenols (1089–1399 μg GAE/g) [2].

The antioxidant activity of formulations (8181 ± 59 μg trolox/g–8421 ± 200 μg trolox/g), did not present statistically significant differences with the control sample (8525 ± 207 μg trolox/g). The antioxidant contribution is due from quinoa (7000–15,000 μg trolox/g) [2].

In vitro digestibility (%) of formulations (70.8 ± 0.1% to 74.6 ± 0.9%), did not present statistically significant differences to control sample (72.54 ± 0.51%). This high digestibility is due to the high er digestibility (83.530 ± 0.007%) of quinoa protein.

The amino acid composition of cookies showed a balanced profile according to the FAO/WHO recommendations (Table 3). The essential amino acids found were histidine (21 ± 6–25 ± 1 mg/g protein), threonine (34 ± 2–41 ± 4 mg/g protein), tyrosine (14 ± 4–20.40 ± 0.07 mg/g protein), phenylalanine (40 ± 2–62 ± 1 mg/g protein), valine (30 ± 3–46 ± 0.2 mg/g protein), isoleucine

(25 ± 5–38.2 ± 0.9 mg/g protein), and tryptophan (3.8 ± 0.6–8.1 ± 0.6 mg/g protein). The cookie formulations with collagen showed a high hydroxyproline content (21 ± 1–48 ± 1 mg/g protein). According to Paul et al. [3], the daily intake of collagen is 2.5–15 g per day, and a 50 g package of Formulations F1 to F4 is within this range.

Table 3. Amino acid profile of cookies with native collagen protein from pota (*Dosidicus gigas*) nape.

Amino Acid	mg Amino Acid/g Protein					
	C	F1	F2	F3	F4	FAO
Aspartic Acid	84.03 ± 0.83 [a]	83.33 ± 15.28 [a]	83.27 ± 35.13 [a]	71.41 ± 2.54 [a]	81.37 ± 3.04 [a]	
Glutamic Acid	142.62 ± 5.38 [a]	131.73 ± 18.35 [a]	136.40 ± 55.34 [a]	109.75 ± 4.84 [a]	120.75 ± 5.16 [a]	
Serine	46.07 ± 3.75 [b]	56.88 ± 1.42 [ab]	68.35 ± 25.19 [ab]	64.90 ± 4.56 [ab]	80.28 ± 1.17 [a]	
Histidine	25.98 ± 1.15 [a]	25.93 ± 0.88 [a]	24.39 ± 6.75 [a]	21.49 ± 6.87 [a]	23.75 ± 0.12 [a]	15
Glycine	57.39 ± 4.12 [b]	115.97 ± 18.31 [ab]	125.54 ± 56.86 [ab]	126.21 ± 1.97 [ab]	158.45 ± 9.03 [a]	
Threonine	40.74 ± 1.67 [a]	41.26 ± 3.61 [a]	41.15 ± 18.12 [a]	34.60 ± 1.87 [a]	39.5 ± 1.39 [a]	23
Arginine	62.11 ± 3.7 [a]	72.13 ± 4.64 [a]	76.36 ± 26.65 [a]	63.65 ± 6.54 [a]	76.94 ± 4.97 [a]	
Alanine	49.33 ± 1.55 [a]	57.88 ± 8.23 [a]	59.30 ± 28.05 [a]	52.06 ± 1.78 [a]	60.85 ± 3.15 [a]	
Proline	71.97 ± 41.97 [a]	108.67 ± 106.99 [a]	138.91 ± 3.06 [a]	113.93 ± 4.25 [a]	130.39 ± 70.83 [a]	
Tyrosine	20.4 ± 0.07 [a]	17.95 ± 1.25 [a]	18.00 ± 5.88 [a]	13.83 ± 3.61 [a]	15.39 ± 0.45 [a]	38 *
Valine	46.04 ± 0.15 [a]	41.31 ± 4.61 [a]	39.23 ± 15.25 [a]	30.11 ± 2.56 [a]	33.04 ± 0.96 [a]	39
Methionine	0.93 ± 0.19 [b]	2.79 ± 1.38 [ab]	5.93 ± 1.31 [a]	4.02 ± 2.35 [ab]	3.03 ± 1.57 [ab]	22 **
Cysteine	6.3 ± 0.07 [a]	4.38 ± 0.22 [ab]	3.81 ± 1.61 [b]	2.12 ± 1.15 [bc]	1.01 ± 0.03 [c]	
Isoleucine	38.2 ± 0.93 [a]	33.68 ± 3.14 [a]	31.70 ± 12.45 [a]	25.21 ± 5.42 [a]	27.24 ± 0.93 [a]	30
Phenylalanine	62.42 ± 1.28 [a]	55.08 ± 5.32 [a]	50.34 ± 19.48 [a]	40.05 ± 2.15 [a]	43.98 ± 1.46 [a]	
Leucine	50.49 ± 1.1 [a]	42.03 ± 3.58 [a]	41.56 ± 16.67 [a]	31.90 ± 3.41 [a]	32.66 ± 2.28 [a]	59
Lysine	43.56 ± 0.81 [a]	35.77 ± 2.87 [ab]	34.93 ± 12.22 [ab]	24.16 ± 2.14 [b]	23.9 ± 0.55 [b]	45
Tryptophan	8.05 ± 0.63 [a]	5.55 ± 0.99 [ab]	4.63 ± 1.06 [b]	4.14 ± 1.51 [b]	3.76 ± 0.61 [b]	6
Hydroxyproline	Nd	21.57 ± 1.34 [d]	31.90 ± 2.21 [c]	40.68 ± 1.72 [b]	48.32 ± 1.24 [a]	

* Tyrosine + Phenylalanine, ** Methionine +Cysteine, Nd: Not determined. Results are expressed as means ± SD (n = 3). [a,b,c,d] values in the same column with different letters differ significantly when $p < 0.05$.

Figure 1 shows the radial graph of the sensory evaluation (color, taste, texture, smell) of the developed cookie formulations, with a positive sensory appreciation equivalent to 77.78% positivity.

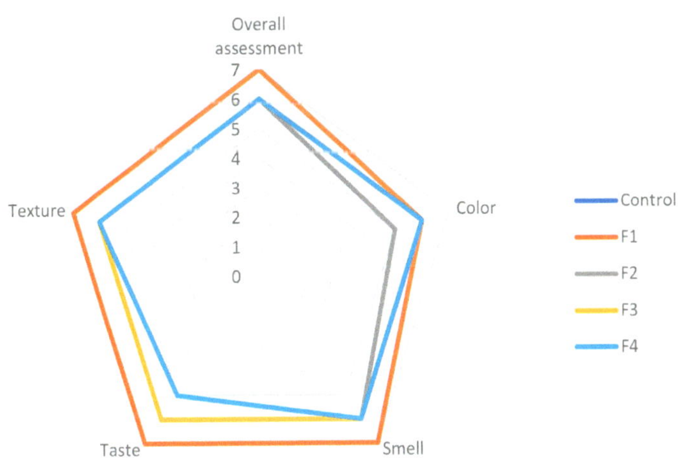

Figure 1. Radial plot of sensory evaluation of cookies with native collagen protein from pota (*Dosidicus gigas*) nape.

The results of microbiological analysis for Formulation F4 (Table 4) were evaluated with the maximum values allowed according to Ministerial Resolution 591-2008-MINSA-Peru [7] and complied with the established requirement.

Table 4. Microbiological analysis, sodium, total sugar, saturated fat, and *trans*-fat content of formulation F4 with native collagen protein from pota (*Dosidicus gigas*) nape.

Analysis	Result	Maximum According to Standard
**Bacillus cereus* Numbering (CFU/g)	<100	10^2 *
E. coli Numbering (MPN/g)	<3	20 *
Fungi: Molds Numbering (CFU/g)	<10	10^3 *
Salmonella Detection	Absence	Absence *
Staphylococcus aureus Numbering (CFU/g)	<10	10^2 *
Total sugars (g/100g)	18.67 ± 1.49	10 **
Sodium (mg/100 g)	343.98 ± 2.75	400 **
Saturated fats (g/100 g)	2.13 ± 0.17	4 **
Trans fats (g/100 g)	<0.01 ± 0.001	5 **
Life time (days)	184	60 ***

Results are expressed as mean ± SD (n = 3). * Ministerial Resolution 591-2008-MINSA-Peru. Sanitary standard that establishes the microbiological criteria of sanitary quality and safety for food and beverages for human consumption. CFU: Colony Forming Units. MPN: Most Probable Number. ** Law No. 30021 Law on the Promotion of Healthy Eating for Children and Adolescents, regulated in Supreme Decree No. 017-2017-SA-Peru. *** De Magalhães et al. [1].

The sodium (343 ± 3 mg/100 g) content, saturated fat (2.1 ± 0.1 g/100 g), and *trans* fat (<0.01 ± 0.001 g/100 g fat) content for Formulation F4 (Table 4) are within the limits established by Peruvian legislation; however, the total sugar (18 ± 1 g/100 g) content is higher due to the high sugar (3–8 g/100 g) content present in quinoa [2].

The developed cookies (F4) presented a shelf life of 184 days by the Rancimat method [5]. According to De Magalhães et al. [1], the cookies lose their physicochemical properties after 60 days of storage due to moisture absorption and decrease in organoleptic quality in the developed product.

Formulation F4 showed important quantity of minerals (Table 5): calcium (3892 ± 194 mg/kg), potassium (3222 ± 161 mg/kg), magnesium (2108 ± 105 mg/kg), zinc (45 ± 2 mg/kg), copper (3.3 ± 0.2 mg/kg), and iron (39 ± 2 mg/kg). According to the daily requirements for children under 5 years of National Institute of Health [10], Formulation F4 covers about 19.46% of daily calcium requirement, 81.08% of magnesium, 7.01% of potassium, and 19.41% of iron.

Table 5. Mineral analysis of Formulation F4 with native collagen protein from pota (*Dosidicus gigas*) nape.

Mineral/Metal	Result
Calcium (mg/kg)	3892.75 ± 194.64
Potassium (mg/kg)	3222.35 ± 161.12
Magnesium (mg/kg)	2108.15 ± 105.41
Zinc (mg/kg)	45.32 ± 2.27
Iron (mg/kg)	38.81 ± 1.94
Cupper (mg/kg)	3.31 ± 0.17
Phosphorus (mg/kg)	<20

Results are expressed as means ± SD (n = 3).

The developed cookies had better nutritional characteristics than commercial cookies. So, the cookie developed could contribute to the reduction in malnutrition in our country.

4. Conclusions

The native collagen from the nape of the squid (*Dosidicus gigas*) can be used to develop functional foods such as cookies. Formulation F4 presented high protein, in vitro digestibility, total polyphenol, and antioxidant activity. Also, Formulation F4 contains high hydroxyproline content and a balanced amino acid profile according to FAO. Formulation F4 had high sensory acceptance and complied with microbiological and health law requirements according to Peruvian legislation. In addition, Formulation F4 had a long shelf life with significant content of calcium, potassium, and magnesium.

Author Contributions: Conceptualization, N.C., A.S., M.T. and R.A.; methodology, N.C., A.S., M.T. and R.A.; investigation and data analysis, N.C., A.S., M.T. and R.A.; writing—original draft preparation, N.C., A.S., M.T. and R.A.; writing—review and editing, N.C., A.S., M.T. and R.A. All authors have read and agreed to the published version of the manuscript.

Funding: This research received no external funding.

Institutional Review Board Statement: Not applicable.

Informed Consent Statement: Not applicable.

Data Availability Statement: Data are contained within the manuscript.

Acknowledgments: This work was developed at Laboratorio de Alimentos Funcionales of Carrera de Ingenieria Industrial and supported by Universidad de Lima, Peru.

Conflicts of Interest: The authors declare no conflicts of interest.

References

1. De Magalhães, C.S.; de Almeida Marques, G.; Bazán-Colque, R.J.; Moraes, E.A.; da Silva EM, M.; Ascheri, J.L.R. The Addition of Pinto Bean Flour and Margarine in the Development of Red Rice-Based Novel Gluten-Free Cookies to Improve the Technological, Sensory and Physicochemical Properties. *Plant Foods Hum. Nutr.* **2023**, *78*, 100–108. [CrossRef] [PubMed]
2. Repo-Carrasco-Valencia, R.; Vidaurre-Ruiz, J.M. Quinoa, Kañiwa, Amaranth, and Lupin as Ingredients in Gluten-Free Baking. In *Native Crops in Latin America*; CRC Press: Boca Raton, FL, USA, 2022; pp. 35–59. [CrossRef]
3. Paul, C.; Leser, S.; Oesser, S. Significant Amounts of Functional Collagen Peptides Can Be Incorporated in the Diet While Maintaining Indispensable Amino Acid Balance. *Nutrients* **2019**, *11*, 1079. [CrossRef] [PubMed]
4. Agüero Correa, A.A. Fuerza 7: Mezcla de Harina de Siete Semillas Ancestrales. 2022. Available online: https://hdl.handle.net/20.500.12724/16465 (accessed on 1 July 2024).
5. Chasquibol, N.; Alarcón, R.; Gonzáles, B.F.; Sotelo, A.; Landoni, L.; Gallardo, G.; García, B.; Pérez-Camino, M.C. Design of Functional Powdered Beverages Containing Co-Microcapsules of Sacha Inchi *P. huayllabambana* Oil and Antioxidant Extracts of Camu Camu and Mango Skins. *Antioxidants* **2022**, *11*, 1420. [CrossRef] [PubMed]
6. Nagai, T.; Yamashita, E.; Taniguchi, K.; Kanamori, N.; Suzuki, N. Isolation and characterisation of collagen from the outer skin waste material of cuttlefish (*Sepia lycidas*). *Food Chem.* **2001**, *72*, 425–429. [CrossRef]
7. Chasquibol, N.; Sotelo, A.; Alarcón, R. Development of Powdered Beverage with Cushuro (*Nostoc commune*) Concentrated Protein and Quinoa (*Chenopodium quinoa*). *Biol. Life Sci. Forum* **2023**, *25*, 2. [CrossRef]
8. Yust, M.M.; Pedroche, J.; Girón-Calle, J.; Vioque, J.; Millán, F.; Alaiz, M. Determination of tryptophan by high-performance liquid chromatography of alkaline hydrolysates with spectrophotometric detection. *Food Chem.* **2004**, *85*, 317–320. [CrossRef]
9. Tapia, M.; Marimón, S.J.; Salazar, N. Development of Extruded Snacks with Protein Hydrolysed from Jumbo Squid (*Dosidicus gigas*) by-Product and Cañihua (*Chenopodium pallidicaule Aellen*). *Biol. Life Sci. Forum* **2023**, *25*, 4. [CrossRef]
10. National Institutes of Health. Hierro. Hoja Informativa Para Consumidores. Available online: https://ods.od.nih.gov/factsheets/Iron-DatosEnEspanol/ (accessed on 1 July 2024).

biology and life sciences forum

MDPI

Proceeding Paper

Development and Characterization of Andean Pseudocereal Bars Enriched with Native Collagen from Pota (*Dosidicus gigas*) By-Products †

Yeromi Coral Paredes Llosa [1], Klidem Francisco Velazquez Carlier [1] and Nancy Chasquibol [2,*]

1 Carrera de Ingeniería Industrial, Universidad de Lima, Av. Javier Prado Este 4600, Fundo Monterrico Chico, Surco, Lima 15023, Peru; 20193214@ulima.edu.pe (Y.C.P.L.); 20194719@ulima.edu.pe (K.F.V.C.)

2 Ingeniería Industrial, Grupo de Investigación en Alimentos Funcionales, Instituto de Investigación Científica, Universidad de Lima, Av. Javier Prado Este 4600, Fundo Monterrico Chico, Surco, Lima 15023, Peru

* Correspondence: nchasquibol@ulima.edu.pe

† Presented at the VI International Congress la ValSe-Food, Lima, Peru, 23–25 September 2024.

Abstract: In recent years, consumers have been increasingly concerned about their health. Therefore, the snack market is rapidly developing more innovative and functional products such as cereal bars. Quinoa (*Chenopodium quinoa*) and kiwicha (*Amaranthus caudatus*) are Andean pseudocereals with protein (10.90–11.35%) content and other functional components that reduce the risk of cardiovascular diseases and inflammatory illnesses. Peru is the world's second largest exporter of Pota (*Dosidicus gigas*), with 476.5 million metric tons in 2023; however, only between 50% and 70% of it has been taken advantage of. Pota by-products such as skins, viscera, and necks have significant protein content (70%) and are discarded. In this investigation, cereal bar formulations with Pota by-products and Andean pseudocereals were optimized and characterized using a five-run simplex centroid mixture design. The effects of two independent variables were examined, namely collagen (2–8%) and binders (22–28%), on the sugar (%), protein (%), and antioxidant (μg Trolox/g dry weight, dw) content as response variables. The optimized cereal bar (M6) showed high protein (21.27 ± 1.51%) content, moisture (10.37 ± 0.04%), ash (2.57 ± 0.03%), fat (15.12 ± 0.15%), carbohydrates (53.67 ± 1.70%), total polyphenol (1570 ± 267 μg Gallic acid equivalent/g dw) content, antioxidant activity (1656 ± 77 μg Trolox/g dw), essential amino acid–leucine (15.65 ± 1.83 mg/g protein) content, and higher in vitro digestibility (78.78 ± 1.40%) than the control sample. The cereal bar had a positive sensory acceptability (88.89%) and complied with Peruvian standards. The functional bar emerges as a nutritious alternative in the food industry and proposes a sustainable solution using Pota by-products, fostering a circular economy.

Keywords: circular economy; functional bars; native collagen; non-communicable diseases; Pota by-products; pseudocereals

1. Introduction

The reuse of fishery waste and by-products could be valuable for developing products with high nutritional content, offering various potential applications across industries and contributing to economic growth. These by-products are rich in proteins, with collagen being the most prevalent, constituting about 30% of the total protein content in animals and found in tissues such as skin, tendons, bones, cartilage, and ligaments [1,2].

Consumer habits are evolving, with more people opting for healthier snacks like nutrient-rich bars that offer real health benefits. Cereal bars have become a popular choice worldwide because they are not only nutritious and low in calories but also help satisfy hunger quickly and support better overall health [3].

This study focuses on the optimization and characterization of the first Andean pseudocereal bar formulated with quinoa (*Chenopodium quinoa*) and kiwicha (*Amaranthus caudatus*), enriched with collagen extracted from by-products of Pota (*Dosidicus gigas*).

2. Materials and Methods

2.1. Raw Material

Flakes of Andean pseudocereals, quinoa (*Chenopodium pallidicaule Aellen*) and kiwicha (*Amaranthus caudatus linnaeus*), along with raisins and pecans, were obtained from a local market in Lima, Peru. Tarwi legume (*Lupinus mutabilis*) was washed and dehydrated in an infrared dryer (IRC D18, Irconfort, Sevilla, Spain) at 60 °C for 12 h in the Functional Foods Laboratory at the Universidad de Lima, Peru. The dehydrated tarwi was then grounded using a food shredder (Grindomix GM200, Restch, Haan, Germany) to obtain tarwi flour. By-products of Pota (*Dosidicus gigas*) were collected from the Piura Department, Peru. The process for extracting collagen from these by-products was determined following the method described by [4] with 5 N NaOH for 30 h, followed by washing, dehydration, grinding in the food shredder, and sieving to 212 µm. The binder was obtained by extracting pineapple juice and mixing it in a 1:20 ratio with flaxseed at 40 °C for 2 h. All samples were stored in aluminized bags at room temperature for subsequent analysis.

2.2. Formulation Optimization

A simplex-centroid mixture design (SCMD) based on response surface methodology (RSM) was implemented. The design included 5 treatments and 2 central point replicates with two components, which were collagen (2%, 3.5%, 5%, 6.5%, 8%) and the pineapple binder (28%, 26.5%, 25%, 23.5%, 22%). This design was based on a cereal bar formulation to evaluate the impact of native collagen derived from Pota by-products. The dependent variables assessed were protein content (%), antioxidant activity (%), and sugar content (%). The highest compound desirability was used as a statistical indicator to validate the product design. The analysis used Minitab 19.0 statistical software (Minitab Inc., State College, Palo Alto, PA, USA).

2.3. Sample Preparation

Five samples were prepared with different amounts of Pota nape collagen (2–8%) and binder (22–28%) while maintaining proportions of quinoa (17%), kiwicha (17%), tarwi flour (20%), raisins (7%), pecans (7%), and pineapple fiber (1%). All the dry ingredients were mixed, including the flakes roasted at 60 °C for 10 min and the crushed nuts. The binder was incorporated into the mixture at room temperature and mixed until homogeneous. The mixture was poured into a mold and refrigerated for 15 min. Afterwards, it was wrapped in butter paper and baked at 180 °C for 12 min, obtaining sticks of 20 to 25 g and stored in aluminized bags for later use.

2.4. Physicochemical Characterization

2.4.1. Proximal Composition

The moisture content of the cereal bars was determined at 110 °C at a constant weight. The ash content was determined by an ignition method (550 °C for 72 h). The fat content was determined with hexane for 9 h, and the total protein content was determined as % nitrogen × 6.25 using a Kjeldahl analyzer (UDK 139, VELP, Usmate Velate, Italy) by official methods. Each analysis was conducted in triplicate, and the results were presented as mean values.

2.4.2. Total Phenolic Content (TPC)

The total phenolic content (TPC) of the Andean pseudocereal bars was measured using the Folin–Ciocalteau method [5]. A total of 15 mg of the sample was dissolved in 4.5 mL of methanol and 2.5 mL of Folin–Ciocalteau reagent 2 N, and the mixture was stirred using a vortex for 1 min. After 5 min, 2.5 mL of sodium carbonate solution (20%) was added, and the mixture was left at 40 °C for 30 min. The absorbance was measured at 760 nm with a spectrophotometer (UV 1280 Vis Spectrophotometer Shimadzu, Kyoto, Japan). The results were expressed as µg of Gallic acid equivalents (GAEs) per gram of dry sample. Each analysis was conducted in duplicate as mean values.

2.4.3. Antioxidant Activity

The antioxidant activity of the samples was evaluated using the DPPH method [5] with certain modifications; a total of 15 mg of the samples was resuspended in 4.5 mL of methanol/acetic acid/water (50:8:42, *v*/*v*/*v*), stirred using a vortex for 1 min, and left in a water bath for 20 min at 80 °C. Then, 3.9 mL of 25 ppm 2, 2-diphenyl-1-picrylhydrazyl (DPPH) radical solution (2.5 mg DPPH in 100 mL MeOH) was added, and the samples were left in the dark at 25 °C. The mixture was stirred using a vortex for 1 min and the absorbance was measured at 517 nm by spectrometry (UV 1280 Vis Spectrophotometer Shimadzu, Kyoto, Japan). The results were reported as µg Trolox/g of the dry sample. Each analysis was performed in triplicate and presented as mean values.

2.4.4. In Vitro Protein Digestibility (IVPD)

In vitro protein digestibility was assessed using the method of Tinus et al. (2012) [6] with slight modifications. The results are reported as

$$\text{IVPD (Digestibility (\%)} = 65.66 + 18.10 \times (\text{pH 0 min} - \text{pH 10 min}))$$

2.5. Determination of Amino Acid Profile

The amino acid profile was determined following the method described by Chasquibol et al. [4]. Acid hydrolysis was conducted for all amino acids except tryptophan, which underwent basic hydrolysis. Protein samples (4 mg) were incubated in 4 mL of 6 N HCl at 110 °C for 24 h in nitrogen-sealed tubes. The hydrolyzed samples were then reconstituted in 1 M sodium borate with 0.02% sodium azide at pH 9.0. The samples were then quantified using HPLC (ARC, Waters, Milford, CT, USA) equipped with a C18 reverse phase column with an internal diameter of 150 mm × 3.9 mm. The calibration curves for each amino acid were developed using a mix of the amino acid standards at the same hydrolysis conditions of the samples. Each analysis was performed in triplicate and presented as mean values.

2.6. Sensory Analysis

A panel of 30 individuals (aged 20 to 51) evaluated the sensory attributes of appearance, color, aroma, taste, flavor, texture, and overall acceptability. The evaluation was conducted using a 9-point hedonic scale (1 = extreme dislike, 9 = extreme liking) [7].

3. Results and Discussion

3.1. Formulation Optimization

The results obtained for all samples are presented in Table 1. With a composite desirability of 66.78%, the optimal formulation of the cereal bar features a collagen concentration of 6% and a pineapple binder concentration of 24%. According to the response variables estimated using Minitab 0.19 software, the enriched cereal bar displays a theorical protein content of 21.77%, a theorical antioxidant activity of 1800.10 µg Trolox/g dw, and a theorical sugar content of 5.95%.

Table 1. Results of the response variables for the optimal formulation.

Sample	Protein (%)	Sugar (%)	DPPH (µg Trolox/g dw)
M1 (2% CPBP)	18.10 ± 0.68	8.03 ± 0.72	1420.84 ± 712.78
M2 (3.5% CPBP)	19.88 ± 1.48	5.88 ± 0.28	1557.24 ± 488.51
M3 (5% CPBP)	20.73 ± 1.46	6.10 ± 0.76	1713.52 ± 520.54
M4 (6.5% CPBP)	24.28 ± 0.04	7.16 ± 0.57	1744.08 ± 482.09
M5 (8% CPBP)	24.31 ± 1.28	5.52 ± 0.49	1870.80 ± 449.99

Results are expressed as means ± SD (n = 3).

3.2. Physicochemical Characterization

Table 2 presents the physicochemical characterization of the optimal bar (M6), control sample bar (C), and commercial cereal bars (CCB). The M6 bar exhibited a high protein content (21 ± 2%), significantly exceeding the 10% threshold required to be classified as a high-protein food according to the FAO and the 8.5% minimum standard for cereal-based products according to NTP-PNAEQW [8]. Regarding sugar content, M6 recorded a value of 3.94 ± 0.03%, which is lower than the control value of 9.5% and complies with the maximum allowable level of 10% established by Law No. 30021 [9]. Additionally, the carbohydrate content (53 ± 2%), ash content (2.57 ± 0.03%), and fat content (15.1 ± 0.2%) in M6 were all superior to those of C and CCB [10].

Table 2. Proximate composition of the optimized Andean pseudocereal bar.

Sample	Moisture (%)	Fat (%)	Protein (%)	Ash (%)	Carbohydrate (%)
M6	10.37 ± 0.04	15.12 ± 0.15	21.27 ± 1.51	2.57 ± 0.03	53.67 ± 1.70
C	7	16.5	7.5	1.5	68
CCB	7.9 ± 2.1	13.1 ± 3.8	5.5 ± 1.3	1.1 ± 0.3	66.0 ± 7

Results are expressed as means ± SD (n = 3). Data available from Oliveira-Carrion et al. [10].

3.2.1. Total Phenolic Content (TPC) and Antioxidant Activity

The phenolic content found in M6 was 1569 ± 268 µg GAE/g dw. Additionally, the antioxidant activity measured by DPPH was 1656 ± 77 µg Trolox/g dw. These results were lower than the commercial bars because our primary antioxidant activity is believed to come from the action of serine protease derived from collagen, used in waste valorization processes and the extraction of naphthol esters, acetates, and triacylglycerols. Unlike commercial cereal bars that rely on various sources of antioxidants, these phenolic compounds are particularly noteworthy for their potent antioxidant properties and anti-inflammatory effects [11].

3.2.2. In Vitro Protein Digestibility (IVPD)

The cereal bar presented high in vitro digestibility IVPD (78.78 ± 1.40%), which was higher than the IVPD extruded snack (74.40 ± 0.51%) [7]. These results indicate that the cereal bar could be an important source of protein and polyphenols for consumers.

3.3. Determination of Amino Acid Profile

The amino acid profile composition showed histidine (25.1 ± 0.9 mg/g protein), isoleucine (35 ± 1 mg/g protein), phenylalanine + tyrosine (56.78 ± 2.6 mg/g protein), threonine (42 ± 2 mg/g protein), tryptophan (10.8 ± 2 mg/g protein), and valine (41 ± 3 mg/g protein). However, leucine (57.60 ± 0.52 mg/g protein), lysine (31.6 ± 0.3 mg/g protein), and methionine + cysteine (18.0 ± 0.5 mg/g protein) are limited components. According to the FAO/WHO recommendations for healthy nutrition, the amino acid composition of the M6 bar displays a well-balanced profile.

3.4. Sensory Analysis

It is concluded that both the optimized bar (M6) and the control sample (C) were well received by the panelists. The M6 bar achieved scores above six in all evaluated attributes, with appearance and color standing out with a rating of eight and an overall acceptability of 88.89%.

4. Conclusions

The incorporation of pseudocereals, such as quinoa and kiwicha, and collagen extracted from Pota by-products, significantly enhanced the physical properties of the (M6) cereal bar, including increased protein and antioxidant content and lower sugar and carbohydrate levels, compared to the control sample (C) and commercial cereal bars (CCB). This development, positively evaluated by a sensory panel and compliant with NTP-PNAEQW (2022) and Law No. 30021 (2013), represents an innovative and sustainable approach that enhances market acceptance.

Author Contributions: All authors have contributed equally to this manuscript. Conceptualization, Y.C.P.L., K.F.V.C. and N.C.; Methodology, Y.C.P.L., K.F.V.C. and N.C.; Investigation and Data analysis, Y.C.P.L., K.F.V.C. and N.C.; Writing-original draft preparation, Y.C.P.L., K.F.V.C. and N.C.; Writing-review and editing, Y.C.P.L., K.F.V.C. and N.C. All authors have read and agreed to the published version of the manuscript.

Funding: This research received no external funding.

Institutional Review Board Statement: Not applicable.

Informed Consent Statement: Not applicable.

Data Availability Statement: Data is contained in the manuscript.

Acknowledgments: This work was developed and supported at Laboratorio de Alimentos Funcionales of Carrera de Ingenieria Industrial by Universidad de Lima, Peru.

Conflicts of Interest: The authors declare no conflict of interest.

References

1. Coppola, D.; Lauritano, C.; Palma, G.; Riccio, G.; Ianora, A. Fish waste: From problem to valuable resource. *Mar. Drugs* **2021**, *19*, 116. [CrossRef] [PubMed]
2. Gisbert, V.G.; Benaglia, S.; Uhlig, M.R.; Proksch, R.; Garcia, R. High-speed nanomechanical mapping of the early stages of collagen growth by bimodal force microscopy. *ACS Nano* **2021**, *15*, 1850–1857. [CrossRef] [PubMed]
3. de Sousa, M.F.; Guimarães, R.M.; de Oliveira Araújo, M.; Barcelos, K.R.; Carneiro, N.S.; Lima, D.S.; Dos Santos, D.C.; de Aleluia Batista, K.; Fernandes, K.F.; Lima, M.C.P.M. Characterization of corn (*Zea mays* L.) bran as a new food ingredient for snack bars. *LWT* **2019**, *101*, 812–818. [CrossRef]
4. Chasquibol, N.; Gonzales, B.F.; Alarcón, R.; Sotelo, A.; Márquez-López, J.C.; Rodríguez-Martin, N.M.; del Carmen Millán-Linares, M.; Millán, F.; Pedroche, J. Optimisation and Characterisation of the Protein Hydrolysate of Scallops (*Argopecten purpuratus*) Visceral By-Products. *Foods* **2023**, *12*, 2003. [CrossRef] [PubMed]
5. Chasquibol, N.; Alarcón, R.; Gonzáles, B.F.; Sotelo, A.; Landoni, L.; Gallardo, G.; García, B.; Pérez-Camino, M.C. Design of Functional Powdered Beverages Containing Co-Microcapsules of Sacha Inchi *P. huayllabambana* Oil and Antioxidant Extracts of Camu Camu and Mango Skins. *Antioxidants* **2022**, *11*, 1420. [CrossRef] [PubMed]
6. Tinus, T.; Damour, M.; Van Riel, V.; Sopade, P.A. Particle size-starch–protein digestibility relationships in cowpea (*Vigna unguiculata*). *J. Food Eng.* **2012**, *113*, 254–264. [CrossRef]
7. Tapia, M.; Marimón, S.J.; Salazar, N. Development of extruded snacks with protein hydrolyzed from jumbo squid (*Dosidicus gigas*) by-product and cañihua (*Chenopodium pallidicaule Aellen*). *Biol. Life Sci. Forum* **2023**, *25*, 4. [CrossRef]
8. Ministerio de Desarrollo e Inclusión Social. Especifiaciones Técnicas de Alimentos que Formas Parte de la Prestación del Servicio Alimentario 2022 del Programa Nacional de Alimentación Escolar Qali Warma. 2021. Available online: https://info.qaliwarma.gob.pe/datpub/uop/catalogo/2022/08-BARRA-DE-CEREALES-YO-LEGUMINOSAS.pdf?v=1.0 (accessed on 15 September 2021).
9. Peruvian Legislation of the Healthy Law No. 30021. Law for the Promotion of Healthy Eating for Children and Adolescents. 2013. Available online: https://cdn.www.gob.pe/uploads/document/file/2182647/PDF%20de%20la%20Ley%20de%20promoci%C3%B3n%20de%20la%20alimentaci%C3%B3n%20saludable%20para%20ni%C3%B1os,%20ni%C3%B1as%20y%20adolescentes.pdf?v=1631731718 (accessed on 17 May 2013).
10. Olivera-Carrión, M.; Giacomino, S.M.; Pellegrino, N.; Sambucetti, M.E. Composición y Perfil Nutricional de Barras de Cereales Comerciales. *Actual. Nutr.* **2009**, *10*, 275–284.
11. Ozogul, F.; Cagalj, M.; Šimat, V.; Ozogul, Y.; Tkaczewska, J.; Hassoun, A.; Ait Kaddour, A.; Kuley, E.; Bhojraj Rathod, N.; Phadke, G.G. Recent developments in valorisation of bioactive ingredients in discard/seafood processing by-products. *Trends Food Sci. Technol.* **2021**, *116*, 559–582. [CrossRef]

biology and life sciences forum

MDPI

Proceeding Paper

Development of a Plant-Based Beverage with Tarwi (*Lupinus mutabilis*) Milk, Polysaccharides from Cushuro (*Nostoc sphaericum*), and Blueberry Extract [†]

Santos Calderón * and Renzo Bañón

Carrera de Ingeniería Industrial, Universidad de Lima, Av. Javier Prado Este 4600, Surco, Lima 15023, Peru; 20182295@aloe.ulima.edu.pe

* Correspondence: 20183691@aloe.ulima.edu.pe

† Presented at the VI International Congress la ValSe-Food, Lima, Peru, 23–25 September 2024.

Abstract: Plant-based milk alternatives are a rapidly growing niche in functional beverages, driven by demand from vegans and lactose-intolerant consumers. However, commercial options often have low protein content (<1.5%) and contain additives. Tarwi, native to the Andes, is rich in protein (45%), oils, and essential nutrients, while blueberries offer flavor and health benefits. Cushuro (*Nostoc sphaericum*), is an Andean microalga and contains high protein content and polysaccharides (42%). The objective of this research was to develop a plant-based beverage (PBB) with tarwi (*Lupinus mutabilis*), polysaccharides from cushuro (PC) and blueberry extract (*Vaccinium corymbosum*)(BE), compared with a control sample with carboxymethyl cellulose (CMC), a commercial thickening agent. The beverage was optimized and characterized using a design of rotatable central composite of surface methodology with nine formulations and four replicates in the center point. The effects of three independent variables were examined: tarwi milk (45% to 55%) and polysaccharides from cushuro (0.05% to 0.2%). The variable blueberry extract content was used as the differential factor between these two conditions. The response variables were protein (%) content and viscosity coefficient (mPa·s). The optimized beverage showed high protein (2.72%) content, viscosity coefficient (23.05 mPa·s), °Brix (2.5), pH (4.49), and acceptable sensory attributes using a 1-to-9-point hedonic scale with 67% positive acceptance. This powdered beverage complied with the Peruvian normative NTP 203.111.2021. Thereby, the plant-based beverage could be a nutritious alternative to functional plant-based beverages.

Keywords: blueberry; polysaccharides; microalgae; tarwi milk; plant-based beverage; protein

1. Introduction

The global PBB market is estimated to be worth USD 12.1 to USD 18.5 billion by 2022, and it is projected to reach over USD 24 billion by 2025. Consumers are increasingly adopting a healthy plant-based diet, which is the reason why the food industry is orienting its production towards plant-based products as an alternative to an animal-based diet. Despite the high nutritional value of animal-based foods, the intake of fatty meats can increase the risk of non-communicable diseases (NCDs) [1]. Most vegetable drinks (almond, oat, cashew, etc.) contain less than 1.5% protein and have a low concentration of essential amino acids (35.1% ± 0.2%) [2]. Tarwi (*Lupinus mutabilis*) is a legume cultivated in the Andes of Peru, Bolivia, and Ecuador. It contains protein between 32% and 51.6%, a high oil (13–24%) content, and minerals such as iron, magnesium, and phosphorus [3]. Blueberries (*Vaccinium corymbosum*) are high in anthocyanins, which act as protective antioxidants against oxidative stress and inflammation. Cushuro (*Nostoc sphaericum*) is an Andean microalga that grows at more than 3000 and 4000 m.a.s.l. It is consumed for its high content of fiber, proteins, vitamins, minerals, and polysaccharide compounds [4] that can be used as stabilizers and thickening properties in the beverage and food industry.

The objective of this research was to optimize, characterize and evaluate the PBB with polysaccharide from cushuro (*Nostoc sphaericum*), tarwi milk (*Lupinus mutabilis*) and blueberry extract (*Vaccinium corymbosum*) with the commercial additive CMC.

2. Materials and Methods

2.1. Raw Material

Tarwi (*Lupinus mutabilis*) was obtained from the Collahuasi district (3310 m.a.s.l.), Recuay province, Ancash region, Peru. The tarwi seeds were placed in a container filled with water (1:2 *w*/*v*). It was soaked for 24 h, changing the water approximately every 6 h. The hydrated tarwi seeds were boiled in water for 2 to 3 h to remove the bitter alkaloids present in the seeds. Blueberry (*Vaccinium corymbosum*) was obtained in the local market of Lima City, Perú.

The polysaccharide from cushuro (*Nostoc sphaericum*) was obtained at Laboratorio de Alimentos Funcionales of the Universidad de Lima, according to Chasquibol et al. [5]. The cushuro flour was mixed with water, heated to 80 °C, and stirred for 30 min. The supernatants were filtered under vacuum using a muslin cloth. The final filtrate was concentrated with a rotary evaporator (Buchi B-100, Flawil, Switzerland) and then precipitated using isopropanol (70%). The precipitate was dried at 50 °C and stored in aluminized bags at room temperature until use.

2.2. Experimental Design

To optimize protein content (%) and to obtain a viscosity coefficient greater than 22 mPa·s, the experimental design employed a response surface methodology (RSM) with a rotating central composite design (CCD) using Minitab 19 software (USA). Two independent variables were used: X1—tarwi milk (TM) (%), and X2— polysaccharides from cushuro (PC) (%). The blueberry extract content variable was determined by the differential between the two previous factors (Table 1). The TM concentration was optimized in the range between 45% and 55% to maintain the purple color of the blueberry and avoid the white color of the tarwi milk. Additionally, the blueberry extract concentration variable was determined by the differential between the above factors (Table 1).

Table 1. Experimental design to development of plant-based beverages formulations with tarwi milk (TM), blueberry extract (BE), and polysaccharides from cushuro (PC).

Formulation	StdOrder	RunOrder	PtType	Blocks	TM (%)	PC (%)	BE (%)
F1	1	1	1	1	45.00	0.05	54.95
F2	2	2	1	1	55.00	0.05	44.95
F3	3	3	1	1	45.00	0.20	54.80
F4	4	4	1	1	55.00	0.20	44.80
F5	5	5	−1	1	42.93	0.13	56.94
F6	6	6	−1	1	57.07	0.13	42.80
F7	7	7	−1	1	50.00	0.02	49.98
F8	8	8	−1	1	50.00	0.23	49.77
F9 [a]	9	9	0	1	50.00	0.13	49.87
F10	10	10	0	1	50.00	0.13	49.87
F11	11	11	0	1	50.00	0.13	49.87
F12	12	12	0	1	50.00	0.13	49.87
F13	13	13	0	1	50.00	0.13	49.87

[a] F9, central point worked with four repetitions. Variable blueberry extract are the difference between tarwi milk (TM) and polysaccharides from cushuro (PC).

2.3. Development of the Beverages

For the development of the plant-based beverages (PBBs), fresh tarwi was liquefied in a solid-to-water ratio 1:3 (*w*/*v*). Blueberries were heated in a beaker with water (1:3 *w*/*v*) at 80 °C for 15 min to retard all biochemical reactions produced by the polyphenol oxidase enzyme during processing. The blueberry extract was vacuum filtered and mixed with the tarwi milk and polysaccharide from cushuro. Complete homogenization was carried out in a blender at a speed of 2900 rpm for 1.5 min and the PBB was pasteurized at 80 °C for 10 min and cooled in a water bath at 20 °C. The control sample was prepared with carboxymethylcellulose (CMC) in substitution of polysaccharide from cushuro.

2.4. Proximal Composition

The ash content was measured using the ignition method at 500 °C for 22 h. The total protein content was calculated as nitrogen percentage multiplied by 6.25 using a Kjeldahl analyzer (UDK 139, VELP, Usmate Velate, Italy) according to official methods [4].

2.5. Viscosity Coefficient

The oscillatory rheological tests were determined by the methodology described by Jeong et al. [6], with some modifications to suit the experimental conditions. The measurements were obtained using a Rheometer (MCR 92 Viscometer, Anton Paar, Graz, Austria) equipped with a plate-plate geometry setup (35 mm diameter and 1 mm gap). The viscosity was determined at 25 ± 0.1 °C and applying a constant shear stress of 20 Pa.

2.6. Antioxidant Activity

The antioxidant activity of the samples was assessed using the DPPH method of Chasquibol et al. [5], described with some modifications, at 517 nm using a UV-Vis spectrophotometer (UV-1280, Shimadzu, Kyoto, Japan). The results were expressed in µg of Trolox per gram of beverage. Each analysis was performed in triplicate, and the values are reported as means.

2.7. Determination of Amino Acid Profile

The amino acid profile was determined following the method of Alaiz et al. [7] with slight modifications. The analysis was performed using HPLC (ARC, Waters, Milford, CT, USA) equipped with a C18 reverse phase column (150 mm × 9 mm internal diameter). Tryptophan was quantified separately by HPLC after undergoing basic hydrolysis. All measurements were conducted in triplicate.

2.8. Sensory Analysis

A sensory evaluation panel composed of 40 individuals, specifically university students aged from 18 to 25, assessed the attributes of appearance, flavor, texture, and overall acceptability. This evaluation employed a 9-point hedonic scale ranging from 1 (extreme dislike) to 9 (extreme like). Panelists were provided with randomly assigned samples. The overall acceptability of the plant-based beverage was determined by calculating the average scores of all sensory attributes.

2.9. Statistical Analysis

The results were presented as mean ± standard deviation, with most measurements performed in triplicate. However, the antioxidant activity measurement was conducted in duplicate. Data were analyzed using analysis of variance (ANOVA) followed by a post hoc Tukey test to identify significant differences ($p < 0.05$) between the two variables, utilizing Minitab® statistical 19 software (State College, PA, USA).

3. Results and Discussion

3.1. Design of Experiment

To optimize protein (%) content was employed the response surface experimental design (RSM) with a rotating central composite (CCD). Table 2 shows thirteen formulations with four replicates at the central point, resulting in protein (%) content and viscosity (mPa·s) coefficients. All the results showed an interaction between tarwi milk (TM), polysaccharides from cushuro (PC), and blueberry extract (BE). The protein content ranged from 1.43% to 2.80%, while the viscosity coefficient varied between 20.52 mPa·s and 41.67

mPa·s, respectively. The highest protein content was obtained for formulation F6 (2.80% ± 0.03) and the highest viscosity coefficient was observed for formulation F8 (41.67 ± 0.04 mPa·s.).

Table 2. Results of experimental design of the plant-based beverage formulations according to tarwi milk (TM), blueberry extract (BE), and polysaccharides from cushuro (PC).

Formulation	TM(%)	PC (%)	BE (%)	Protein (%)	VC (mPa·s)
F1	45.00	0.05	54.95	1.43 ± 0.04	23.32 ± 0.07
F2	55.00	0.05	44.95	2.57 ± 0.02	23.44 ± 0.09
F3	45.00	0.20	54.80	1.46 ± 0.03	37.58 ± 0.12
F4	55.00	0.20	44.80	2.49 ± 0.01	39.56 ± 0.06
F5	42.93	0.13	56.94	1.59 ± 0.02	29.05 ± 0.03
F6	57.07	0.13	42.80	2.80 ± 0.03	31.31 ± 0.05
F7	50.00	0.02	49.98	2.09 ± 0.02	20.52 ± 0.08
F8	50.00	0.23	49.77	2.08 ± 0.02	41.67 ± 0.04
F9	50.00	0.13	49.87	2.05 ± 0.01	32.56 ± 0.02
F10	50.00	0.13	49.87	2.10 ± 0.03	32.33 ± 0.01
F11	50.00	0.13	49.87	2.14 ± 0.02	31.42 ± 0.12
F12	50.00	0.13	49.87	2.16 ± 0.04	32.43 ± 0.07
F13	50.00	0.13	49.87	2.17 ± 0.02	35.30 ± 0.05

Results are expressed as means ± SD (n = 3).

Table 3 shows the regression models and adjustments of the central composite design (CCD) for the independent variables: (TM) and (PC) based on the dependent variables (protein content and viscosity coefficient). After obtaining the results of the output variables, Minitab was used to optimize the plant-based beverage formulation. The parameters were set according to the results of TM (45–55%) content and PC (0.05–0.2%) content in order not to increase the tarwi milk content more than 55%, which could interfere with the purple color of the blueberry extract. The protein content was set a maximum value of 2.8%, while the viscosity coefficient was established a maximum value of 25 mPa-s.

Table 3. Regression models of the central composite design (CCD) for independent and dependent variables evaluated in the formulations of plant-based beverages.

Variable	Equation
Protein content	$y = -3.60 + 11.7\ TM + 612\ PC - 1.1\ TM \times TM - 102444\ PC \times PC - 733\ TM \times PC$ $R^2 = 95.85\%$ p-value = 0.000
Viscosity coefficient	$y = -103.7 + 490\ TM + 7294\ PC - 7294\ PC - 492\ TM \times TM - 1375447\ PC \times PC + 12382\ TM \times PC$ $R^2 = 98.85\%$ p-value = 0.000

Figure 1 shows that protein content increase in relation to the percentage of TM content. The region of greatest interest for the protein content of more than 2% was upper than 50%. The viscosity coefficient increased in direct relation to the concentration of the polysaccharide from cushuro, in the region of major interest superior to 0.01%, for a viscosity coefficient above 20 mPa-s.

Figure 2 shows the optimization of the independent variables that affect to protein content and viscosity capacity. The optimal formulation was: tarwi milk (56.98%), polysaccharide from cushuro (0.07%) and blueberry extract (42.95%), with protein content of 2.8% and viscosity coefficient of 25 mPa·s.

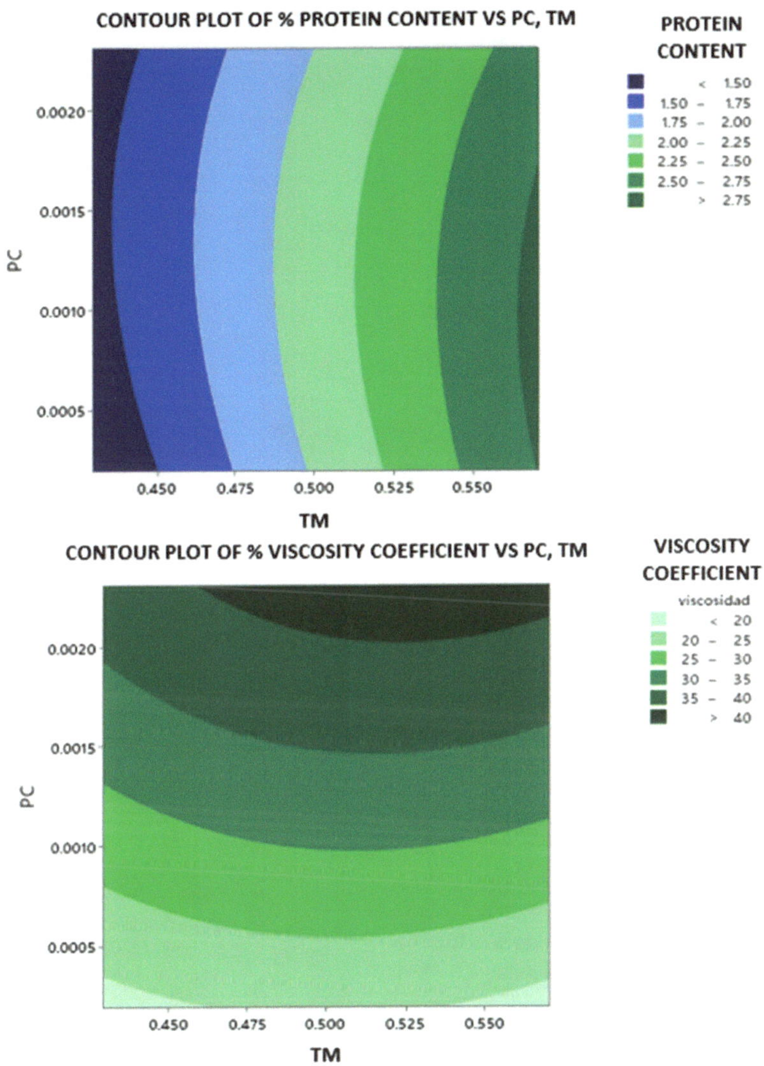

Figure 1. Contour plot of the central composite design for protein content and viscosity coefficient.

3.2. Physicochemical Characterization

Table 4 shows the physicochemical characterization of the PBBs. The PBB with PC showed a higher protein (2.72% $\pm$ 0.12) content than the PBB with CMC (2.36 $\pm$ 0.15). Lopes et al. [8] showed protein content between 1.8% and 2.4% (w/v) in the tarwi-based beverage. On the other hand, seed-based beverages are characterized by their variable lipid (ca. 1–7%) and protein (ca. 1–5%) content [9].

The viscosity coefficient in the PBB with PC was much higher (23.05 $\pm$ 0.09 mPa·s) than the PBB with CMC (21.04 $\pm$ 0.13 mPa·s) (Table 4). PC is a branched hydrocolloid that can form many hydrogen bonds and greatly increase the viscosity. In addition, a significant interaction effect of PC concentration on the °Brix was observed.

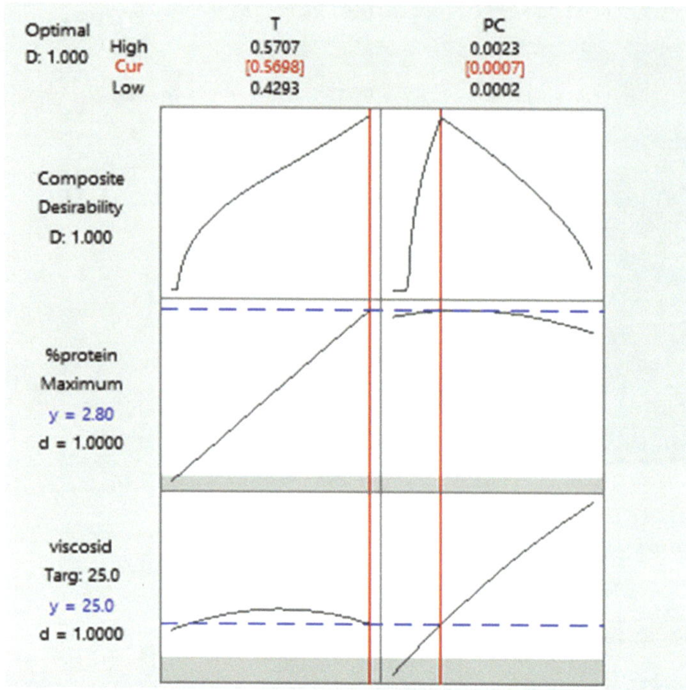

Figure 2. Optimization of the Box–Behnken design obtained from the Minitab 19 software.

Table 4. Physicochemical characterization of plant-based beverage with polysaccharide from cushuro (PBB+PC) and plant-based beverage with carboxymethyl cellulose (PBB+CMC).

Sample	Protein Content (%)	Viscosity Coefficient (mPa·s)	pH	°Brix	Ash (%)	DPPH (µg Trolox/g Beverage)
PBB + PC	2.72 ± 0.12 [a]	23.05 ± 0.09 [a]	4.49 ± 0.2 [a]	2.5 ± 0.24 [a]	0.03 ± 0.01 [a]	1098.98 ± 63.48 [a]
PBB + CMC	2.36 ± 0.15 [b]	21.04 ± 0.13 [b]	4.3 ± 0.13 [a]	2.3 ± 0.31 [a]	0.02 ± 0.02 [a]	746.44 ± 32.98 [b]

Results are expressed as means ± SD (n = 3). a and b values in the same column with different letters differ significantly when $p < 0.05$. PBB + PC: plant-based beverage with polysacharides from cushuro; PBB + CMC: plant-based beverage with carboxymethylcellulose.

The addition of hydrocolloids (PC and CMC) (0.07% *w/v*) and BE (42.95% *w/v*) also influenced pH. The pH (4.49 ± 0.2) of PBB + PC was higher than the pH (4.3 ± 0.13) of PBB + CMC. The addition of commercial CMC (0.1, 0.3 and 0.5%) to snake fruit syrup did not affect pH. However, the addition of pectin (0.2 and 0.3%) and xanthan gum (0.1 and 0.2%) to orange juice with pulp showed a significant influence on pH [10].

The DPPH results are shown in Table 4. The PBB with PC had a significantly higher DPPH (1098.98 µg GAE/g ± 63.48) value than the PBB with CMC (746.44 µg GAE/g ± 32.98). The observed difference in DPPH values was due to the contribution of antioxidants from the cushuro polysaccharide and blueberry extract.

3.3. Amino Acid Profile

According to Table 5, the relevant amino acids in the PBB with PC were lysine (58.08 ± 12.71 mg/g protein), threonine (56.15 ± 35.89 mg/g protein) and tyrosine (45.43 ± 14.59 mg/g protein). According to the protein of reference established by FAO/WHO [4], the composition of amino acids in the plant-based beverage presented a balanced profile. Walther et al. [2] reported that plant-based beverages made from oats, coconut, cashew, spelt and other ingredients had an average value of the sum of amino acids (AA) ranging from 8.1 mg AA/g protein to 11.5 mg AA/g protein. The PBB with PC had a high percentage of tryptophan, above the recommendations of the FAO (6 mg/g protein).

Table 5. Amino acid profile of plant-based beverages with polysaccharide from cushuro (PBB+PC) and plant-based beverages with carboxymethyl cellulose (PBB+CMC).

Amino Acids	PBB + PC (mg AA/g Protein)	PBB + CMC (mg AA/g Protein)	FAO/WHO (mg AA/g Protein)
Aspartic acid	24.19 ± 1.51 ^{a}	22.36 ± 1.79 ^{a}	
Glutamic acid	35.14 ± 15.00 ^{a}	33.63 ± 1.10 ^{a}	
Serine	44.23 ± 1.23 ^{a}	44.52 ± 3.79 ^{a}	
Histidine	22.11 ± 2.00 ^{a}	33.45 ± 4.37 ^{b}	15
Glycine	20.00 ± 10.22 ^{a}	18.15 ± 2.90 ^{a}	
Threonine	56.15 ± 35.89 ^{a}	55.65 ± 4.04 ^{a}	23
Arginine	16.99 ± 7.04 ^{a}	13.97 ± 1.77 ^{a}	
Alanine	14.51 ± 1.82 ^{a}	16.04 ± 4.36 ^{a}	
Proline	33.76 ± 21.18 ^{a}	32.85 ± 2.35 ^{a}	
Tyrosine	45.43 ± 14.59 ^{a}	45.61 ± 13.43 ^{a}	38
NH3	34.30 ± 4.59 ^{a}	34.56 ± 1.54 ^{a}	
Valine	32.22 ± 14.27 ^{a}	32.97 ± 12.73 ^{a}	39
Methionine	32.17 ± 13.13 ^{a}	32.16± 13.24 ^{a}	22
Cysteine	25.78 ± 17.01 ^{a}	26.71 ± 4.70 ^{a}	
Isoleucine	17.14 ± 1.23 ^{a}	44.80 ± 4.24 ^{b}	30
Leucine	42.33 ± 4.69 ^{a}	28.37 ± 2.77 ^{b}	59
Phenylalanine	45.09 ± 11.18 ^{a}	35.05 ± 4.61 ^{b}	
Lysine	58.08 ± 12.71 ^{a}	56.42 ± 5.90 ^{a}	45
Tryptophan	15.03 ± 3.57 ^{a}	13.26 ± 6.47 ^{a}	6

Results are expressed as means ± SD (n = 3). a and b values in the same row with different letters differ significantly when $p < 0.05$. Note: According to the FAO Database on Protein Quality Evaluation [1].

3.4. Sensory Analysis

A comparative analysis was performed between the sample (PBB + PC) and the control sample (PBB + CMC). Both beverage displayed a light purple coloration. (PBB + PC) attained an average consumer acceptance rate of 67%, whereas (PBB + CMC) received an average approval rating of 62%. Regarding specific sensory attributes, including appearance, colour, flavour, texture, and taste, (PBB + PC) was rated at 8.7, 7.8, 6.5, 6.7, and 7, respectively. (PBB + CMC) obtained scores of 8.5 for appearance, 7.2 for colour, 6.6 for flavour, 5.9 for texture, and 6.8 for taste.

4. Conclusions

The PBB with PC was optimized using a central composite rotational design and validated experimentally. The formulation optimized was: tarwi milk (56.98%), polysaccharide from cushuro (0.07%), and blueberry extracts (42.95%). The synergy between tarwi milk, polysaccharide from cushuro and blueberry extract provided a unique nutritional profile with high protein content, high viscosity, °Brix, antioxidant capacity and amino acid profile than control sample with CMC. The favorable sensory acceptance suggests strong potential for market success, positioning this beverage as an innovative alternative in the plant-based and functional food sector.

Author Contributions: All authors have contributed equally to this manuscript. Conceptualization, S.C. and R.B.; Methodology, S.C. and R.B.; Investigation and Data analysis, S.C. and R.B.; Writing—original draft preparation, S.C. and R.B.; Writing—review and editing, S.C. and R.B. All authors have read and agreed to the published version of the manuscript.

Funding: This research received no external funding.

Institutional Review Board Statement: Not applicable.

Informed Consent Statement: Not applicable.

Data Availability Statement: Data is contained within the manuscript.

Acknowledgments: This work was developed in the Laboratorio de Alimentos Funcionales of Carrera de Ingeniería Industrial and supported by the Universidad de Lima, Peru.

Conflicts of Interest: The authors declare no conflict of interest.

References

1. Xiao, X.; Zou, P.-R.; Hu, F.; Zhu, W.; Wei, Z.-J. Updates on Plant-Based Protein Products as an Alternative to Animal Protein: Technology, Properties, and Their Health Benefits. *Molecules* **2023**, *28*, 4016. [CrossRef]
2. Walther, B.; Guggisberg, D.; Badertscher, R.; Egger, L.; Portmann, R.; Dubois, S.; Rezzi, S. Comparison of nutritional composition between plant-based drinks and cow's milk. *Front. Nutr.* **2022**, *9*, 988707. [CrossRef] [PubMed]
3. Huaringa-Joaquin, A.; Saldaña, C.L.; Saravia, D.; García-Bendezú, S.; Rodriguez-Grados, P.; Salazar, W.; Camarena, F.; Injante, P.; Arbizu, C.I. Assessment of the Genetic Diversity and Population Structure of the Peruvian Andean Legume, Tarwi (*Lupinus mutabilis*), with High Quality SNPs. *Diversity* **2023**, *15*, 437. [CrossRef]
4. Chasquibol, N.; Sotelo, A.; Alarcón, R. Development of Powdered Beverage with Cushuro (Nostoc commune) Concentrated Protein and Quinoa (*Chenopodium quinoa*). *Biol. Life Sci. Forum.* **2023**, *25*, 2. [CrossRef]
5. Chasquibol, N.; Sotelo, A.; Tapia, M.; Alarcón, R.; Goycoolea, F.; Perez-Camino, M.d.C. Co-Microencapsulation of Cushuro (*Nostoc sphaericum*) Polysaccharide with Sacha Inchi Oil (*Plukenetia huayllabambana*) and Natural Antioxidant Extracts. *Antioxidants* **2024**, *13*, 680. [CrossRef] [PubMed]
6. Jeong, Y.; Lim, W.; Yoo, B. Relationship between Syringe Flow Measurements and Viscosity of Nectar-Thick Beverages for Dysphagia Management. *Foods* **2021**, *10*, 1981. [CrossRef] [PubMed]
7. Alaiz, M.; Navarro, J.L.; Girón, J.; Vioque, E. Amino Acid Analysis by High-Performance Liquid Chromatography after Derivatization with Diethyl Ethoxymethylenemalonate. *J. Chromatogr. A* **1992**, *591*, 181–186. [CrossRef] [PubMed]
8. Lopes, M.; Pierrepont, C.; Duarte, C.M.; Filipe, A.; Medronho, B.; Sousa, I. Legume Beverages from Chickpea and Lupin, as New Milk Alternatives. *Foods* **2020**, *9*, 1458. [CrossRef] [PubMed]
9. Fructuoso, I.; Romão, B.; Han, H.; Raposo, A.; Ariza-Montes, A.; Araya-Castillo, L.; Zandonadi, R.P. An overview on nutritional aspects of plant-based beverages used as substitutes for mow's milk. *Nutrients* **2021**, *13*, 2650. [CrossRef] [PubMed]
10. Mohammadi, M.; Mahdavi-Yekta, M.; Reihani, S.F.S.; Khorshidian, N.; Habibi, M.; Khaneghah, A.M. Assessment of Physicochemical Properties of Orange Juice Concentrate Formulated with Pectin, Xanthan, and CMC Hydrocolloids. *Int. J. Food Sci.* **2024**, *2024*, 7013553. [CrossRef] [PubMed]

biology and life sciences forum

MDPI

Proceeding Paper

Design of a Functional Mayonnaise Enriched with Omega-3 from Sacha Inchi (*Plukenetia huayllabambana*) Oil and Chia (*Salvia hispanica* L.) Mucilage N.I. One †

Mateo Micali * and Angel Valcarcel

Carrera de Ingeniería Industrial, Universidad de Lima, Av. Javier Prado Este 4600, Fundo Monterico Chico, Surco, Lima 15023, Peru; 20163680@aloe.ulima.edu.pe

* Correspondence: 20172898@aloe.ulima.edu.pe

† Presented at the VI International Congress la ValSe-Food, Lima, Peru, 23–25 September 2024.

Abstract: Sacha Inchi seeds (*Plukenetia huayllabambana*) are highly regarded for their nutritional richness, specifically their high omega-3 content. Chia seed (*Salvia hispanica* L.) mucilage is recognized for its emulsion abilities. There is growing demand for innovative mayonnaise formulations using healthier, plant-based alternatives. This study developed a plant-based mayonnaise (PBM) by replacing egg yolks with chia seed mucilage (CSM) and using Sacha Inchi seed oil (SIO), achieving sensory qualities similar to traditional mayonnaise. Five formulations of PBM were evaluated, with variations in CSM content (1% to 3%) and water content (43% to 45%) and using salt (0.5%), oil (48%), pepper (0.5%) and lemon juice (5%). PBM was evaluated based on omega-3 (%) content, total fat (%) content, stability of emulsion (%), rheology and physicochemical properties. Formulation with 3% of CSM was the optimal option due to its emulsion stability (98.56%) and rheology, very similar to those of traditional mayonnaise (99.13%). PBM formulation with 3% CSM showed the highest omega-3 fatty acid content of 55.36% for 100 g fat, compared with the 0.27% found in traditional mayonnaise. The PBM formulation with 3% CSM also showed important characteristics such as phenolic content (310.814 µg GAE/g ms), antioxidant activity (1991.79 µg Trolox/g ms), Ph (4.24), a peroxide index (11.92 meq-O_2/Kg oil), an acidity index (3.59 mg KOH/g), a shelf life study and proximal composition. This study underscores the potential of CSM and SIO in mayonnaise formulations, addressing concerns with traditional options.

Keywords: chia; emulsion; mayonnaise; mucilage; sacha inchi; omega-3

1. Introduction

Today's research is aimed at identifying alternative emulsifying agents, particularly plant-based proteins, to meet the demand for egg-free mayonnaise [1]. The quest for healthier and more nutritious mayonnaise formulations has led to the exploration of innovative ingredients such as chia mucilage and Sacha Inchi oil. Chia (*Salvia hispanica* L.) seed mucilage, a complex polysaccharide, holds technological promise regarding its ability to thicken and emulsify, making it a valuable ingredient in food formulations [2]. Similarly, Sacha Inchi (*Plukenetia huayllabambana.*) has garnered attention for its high oil content and superior omega-3 fatty acid composition compared to other varieties [3]. The exploration of Chia mucilage and Sacha Inchi oil as potential ingredients in mayonnaise formulations represents a key step in meeting objectives and catering to evolving consumer preferences for healthier, sustainable food choices.

2. Materials and Methods

2.1. Raw Materials

The materials used for this study included Chia seeds (*Salvia hispanica* L.), Sacha Inchi oil (*Plukenetia huayllabambana* L.), lemons, black pepper, salt, and water. Sacha Inchi seeds were sourced from Rodriguez de Mendoza, Region of Amazonas-Peru, while the other ingredients were purchased locally.

2.2. Extraction of Chia Mucilage and Sacha Inchi Oil

Chia seed mucilage (CSM) extraction was obtained according to the methodology of Fernandes and Salas-Mellado et al. [1], albeit with some modifications. CSM was exuded from whole seeds using distilled water (40:1 water/seed ratio) overnight at 25 °C. The seeds were blended for 30 s, then centrifuged at 2500 rpm for 5 min to separate the mucilage, which was air-dried at 50 °C, ground, and stored at 4 °C. The yield was 9.28% (g mucilage/g chia seed). Sacha Inchi oil was cold-pressed in the Laboratorio de Alimentos Funcionales, Universidad de Lima, and stored at 4 °C in a dark environment until used.

Sample Formulation

Five PBM samples were created with different amounts of CSM and water (Table 1). The control sample used egg yolk as an emulsifier and vegetable oil, following the Peruvian Technic Standard for mayonnaise (NTP 209.033) [4], with 76% oil, 5% lemon juice and 6% egg yolk. CSM levels in PBM formulations ranged from 1% to 3%, while Sacha Inchi oil was constant at 48%.

Table 1. PBM [a] composition.

Materials	Sample 1 (CSM [b] 1%)	Sample 2 (CSM [b] 1.5%)	Sample 3 CSM [b] 2%)	Sample 4 (CSM [b] 2.5%)	Sample 5 (CSM [b] 3%)
% SIO	48	48	48	48	48
% CSM	1	1.5	2	2.5	3
% Salt	0.5	0.5	0.5	0.5	0.5
% Pepper	0.5	0.5	0.5	0.5	0.5
% Water	45	44.5	44	43.5	43
% Lemon	5	5	5	5	5

[a] Plant-based mayonnaise. [b] Chia seed mucilage.

2.3. Development of a Functional Vegan Mayonnaise

PBM formulations followed Cornelia et al. [5] with modifications. CSM was hydrated with water and lemon juice, then mixed with a manual blender. SIO was added in small amounts during mixing. After adding all the oil, salt and pepper were included. To ensure homogeneity, the mixture was processed in the Silverson Mixer L5M-A. ((Silverson Machines, East Longmeadow, MA, USA)).

2.4. Steady Flow Behavior

The steady flow behavior of the mayonnaise was assessed over a shear rate range of 0–100 s^{-1} using a parallel plate configuration with a 1 mm gap between the rheometer probe and the sample plate. To analyze the flow characteristics, the data were used to model the Herschel–Bulkley equation:

$$t = k\,(y^n) + \sigma$$

where "t" is the shear stress (Pa), "y" is the shear rate (1/s), "k" is the consistency index (Pa s), "n" is the flow behavior index, and σ is the yield stress (Pa) [6].

2.5. Emulsion Stability

Emulsion stability was assessed using the methodology outlined by Fernandes and Salas-Mellado et al. [1], albeit with modifications. The samples were stored in 15 mL tubes at 50 °C for two days, then centrifuged at 2500 rpm. The mass of the precipitated fraction was measured, and stability was calculated using the following equation:

$$ES\ (\%) = F1/F0 \times 100$$

where "F1" represents the mass of the precipitated fraction, and "F0" represents the total mass of the sample.

2.6. Proximal Composition

Moisture content was measured at 110 °C until a constant weight. Ash content was determined at 550 °C for 72 h by the ignition method. pH values were recorded at 25 °C with a pH meter (SevenCompact pH meter S220, Mettler Toledo, Singapore). Fat content was assessed with hexane for 9 h, and total protein content was calculated as % nitrogen × 6.25 using a Kjeldahl analyzer (UDK 139, VELP, Usmate Velate, Italy) following official methods.

2.7. Determination of Functional Fatty Acids

Fatty acid profile was obtained as follows: "ISO 12966-1:2014 Animal and Vegetable fats and oils—Gas chromatography of fatty acid methyl esters—Part 2: Preparation of methyl esters of fatty acids"

Phenolic Analysis

The total phenolic content (TPC) of the mayonnaise was determined by the Folin–Ciocalteau method [3] at 760 nm using a UV 1280 Vis Spectrophotometer from Shimadzu, Kyoto, Japan. The results were expressed as µg of gallic acid equivalent (GAE)/g mayonnaise.

2.8. Antioxidants Analysis

The antioxidant activity of samples was assessed using the DPPH method [3], with some modifications, at 517 nm by spectrometry (UV 1280 Vis Spectrophotometer Shimadzu, Kyoto, Japan). The results were expressed as µg Trolox/g mayonnaise.

2.9. Oxidative Stability

Oxidative stability was evaluated using an 892 Professional Rancimat© (Metrohm, Herisau, Switzerland) following Chasquibol et al.'s method [3]. Shelf life at 25 °C was estimated by determining induction periods at 80, 90, and 100 °C, with all measurements performed in triplicate. Additionally, acid and peroxide indices were assessed according to Chasquibol et al. [3].

2.10. Sensory Evaluation

A sensory evaluation was conducted with 45 university students (aged 18–26) to assess appearance, flavor, texture, and overall acceptability using a 9-point hedonic scale (1 = "dislike extremely", 9 = "like extremely"). Panelists received randomized samples in coded plastic cups labeled M1 [3].

2.11. Statistical Analysis

A 2k design with two replications and a central point was conducted, using quantity of chia and water as variables. The viscosity and emulsion stability results were presented as means with standard deviations, with all analyses being performed in triplicate. Data were analyzed using ANOVA at a 95% confidence level (Minitab Inc., State College, Palo Alto, CA, USA).

3. Results

According to the experimental design, it is concluded that Chia, water, and their interaction are significant, yielding an R^2 of 99.89%, R^2 (adj) of 99.80%, and R^2 (pred) of 99.55%, as shown in Figure 1. Viscosity optimization, targeting a viscosity of 2100, resulted in a formulation with 2.93% chia and 43% water (Figure 1).

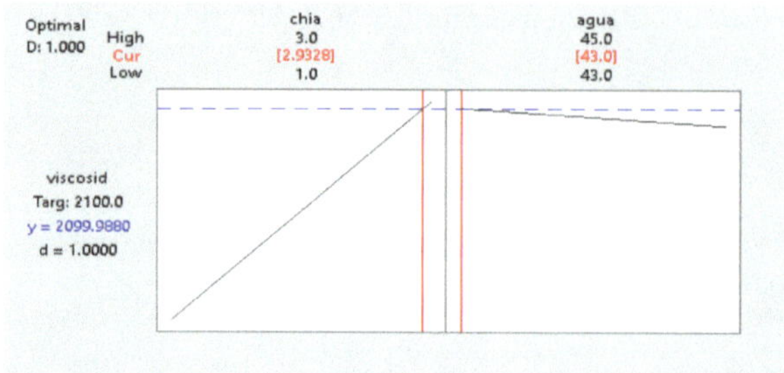

Figure 1. The optimization of the design obtained from Minitab 19.0 Software.

All variables increased with higher proportions of CSM in PBM, with the highest values being observed at 3% CSM: "σ" (53.47 ± 24.12 to 122.05 ± 51.65), "k" (70.31 ± 24.96 to 129.88 ± 51.92), and "n" (0.64 ± 0.01 to 0.37 ± 0.01%). Flow behavior curves are shown in Figure 2.

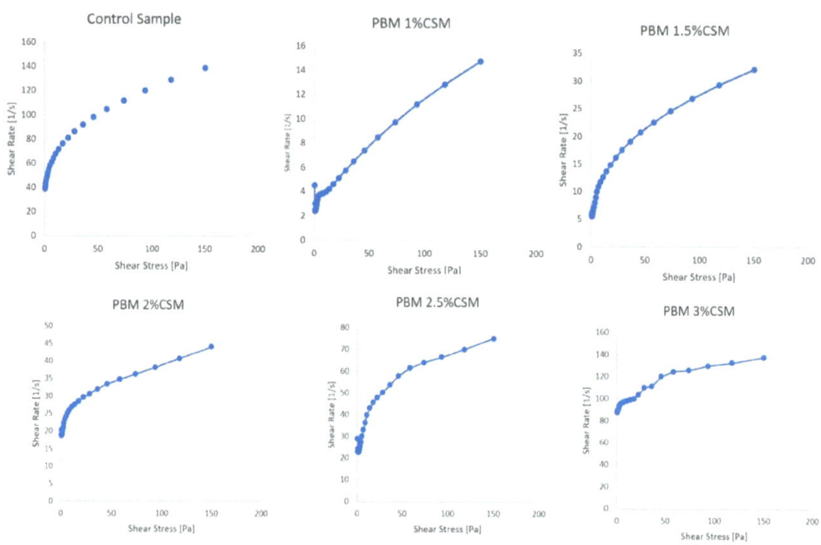

Figure 2. Flow Behavior of the mayonnaise samples.

Emulsion stability in Table 2 shows a direct correlation with increasing CSM, with lower phase separation at higher mucilage concentrations; similar results were found in Caldeira Soares et al. [7]. PBM with 3% CSM best mimicked traditional mayonnaise properties, and subsequent analyses focused on this sample.

Table 2. Steady flow behavior and emulsion stability of PBM [a].

Result	Control	Sample 1 (CSM [b] 1%)	Sample 2 (CSM [b] 1.5%)	Sample 3 (CSM [b] 2%)	Sample 4 (CSM [b] 2.5%)	Sample 5 (CSM [b] 3%)
Viscosity	2189.75 ± 55.70	156.38 ± 10.83	463.43 ± 27.71	740.92 ± 51.48	1298.85 ± 206.99	2467.871 ± 285.08
ES	99.13 ± 0.08	63.08 ± 5.04	81.22 ± 2.02	92.60 ± 0.97	97.85 ± 1.08	98.57 ± 0.14

[a] Plant-based mayonnaise. [b] Chia seed mucilage. The results are expressed as means ± SD (n = 3).

PBM showed a lower lipid (75.64 ± 0.40% to 48.54 ± 1.61%) content, protein (1.21 ± 0.22% to 1.19 ± 0.15%), and ash (1.8 ± 0.33% to 1.64 ± 0.37%) than the control sample, while carbohydrates (2.58 ± 0.16% to 3.15 ±1.12%) and moisture (18.77 ± 0.20% to 45.49 ± 0.49%)

content were higher than the control sample. PBM showed a higher TPC (256.62 ± 45.44 µg GAE/g to 253.962 ± 29.83 µg GAE/g) content, DPPH antioxidant activity (1991.79 ± 258.48 µg Trolox/g), and pH (4.42 to 3.26) than the control sample. PBM had a lower acid (5.49 ± 0% to 3.59 ± 0%) index and peroxide (7.25 ± 3.70 meq O2/kg to 11.92 ± 5.76 meq O2/kg). PBM was rich in omega-3 fatty acids (55.36%), followed by omega-6 (29.32%) and omega-9 (8.08%), due to sacha inchi (Plukenetia huayllabambana) oil; while the control sample had more omega-6 (50.91%) and omega-9 (36.12%), with a small amount of omega-3 (0.27%). The estimated shelf life for Rancimat® at 25 °C was 715 to 732 h, shorter than that control sample of 3221 to 3433 h, because the PBM does not contain commercial additives. PBM had a slight beige color and no characteristic odor. In sensory evaluation, consumer acceptability scored above 5 on a nine-point hedonic scale, leading to an overall acceptability rate of 71%. Average scores for appearance, flavor, texture, and overall acceptability were 5.24, 6.78, 5.29, and 6.06, respectively.

4. Conclusions

The study shows that 3% chia seed mucilage effectively replicates traditional mayonnaise. Replacing vegetable oils with Sacha Inchi oil enhances nutritional value, with the vegan mayo showing higher phenolic concentration, antioxidant activity, and omega-3 content. It also has less total fat and is well accepted by evaluators, offering a healthier alternative that aligns with current market trends.

Author Contributions: Conceptualization, M.M. and A.V.; methodology, M.M. and A.V.; software, M.M. and A.V.; validation M.M. and A.V.; formal analysis, M.M. and A.V.; investigation, M.M. and A.V.; resources, M.M. and A.V.; data curation, M.M. and A.V.; writing—original draft preparation, M.M. and A.V.; writing—review and editing, M.M. and A.V.; visualization, M.M. and A.V.; supervision, M.M. and A.V.; project administration, M.M. and A.V.; funding acquisition, M.M. and A.V. All authors have read and agreed to the published version of the manuscript.

Funding: This research received no external funding.

Institutional Review Board Statement: Not applicable.

Informed Consent Statement: Not applicable.

Data Availability Statement: Data is contained within the document.

Acknowledgments: This work was supported and developed at Laboratorio de Alimentos Funcionales, Universidad de Lima, Peru.

Conflicts of Interest: The authors declare no conflict of interest.

References

1. Fernandes, S.S.; Salas-Mellado, M.d.l.M. Development of mayonnaise with substitution of oil or egg yolk by the addition of chia (*Salvia hispanica* L.) mucilage. *J. Food Sci.* **2018**, *83*, 74–83. [CrossRef] [PubMed]
2. Capitani, M.I.; Nolasco, S.M.; Tomás, M.C. Stability of oil-in-water (O/W) emulsions with chia (*Salvia hispanica* L.) mucilage. *Food Hydrocoll.* **2016**, *61*, 537–546. [CrossRef]
3. Chasquibol, N.A.; Gallardo, G.; Gómez-Coca, R.B.; Trujillo, D.; Moreda, W.; Pérez-Camino, M.C. Glyceridic and unsaponifiable components of microencapsulated Sacha Inchi (*Plukenetia huayllabambana* L. and *Plukenetia volubilis* L.) edible oils. *Foods* **2019**, *8*, 671. [CrossRef] [PubMed]
4. *NTP 209.033*; Norma Técnica Peruana, Amendment—MAYONESA. Dirección de Normalización—INACAL: Lima, Peru, 2017.
5. Cornelia, M.; Siratantri, T.; Prawita, R. The Utilization of Extract Durian (*Durio zibethinus* L.) Seed Gum as an Emulsifier in Vegan Mayonnaise. *Procedia Food Sci.* **2015**, *3*, 1–18. [CrossRef]

6. Hijazi, T.; Karasu, S.; Tekin-Çakmak, Z.H.; Bozkurt, F. Extraction of Natural Gum from Cold-Pressed Chia Seed, Flaxseed and Rocket Seed Oil By-Product and Application in Low Fat Vegan Mayonnaise. *Foods* **2022**, *11*, 363. [CrossRef] [PubMed]
7. De Castro Caldeira Soares, A.; Tavares, M.B.F.B.; de Paiva Marcos Ortega, E.; Vidigal, M.C.T.R.; de Castro Santana, R. Rheological and sensorial evaluation of fruit nectar with chia mucilage. *Int. J. Gastron. Food Sci.* **2024**, *35*, 100849. [CrossRef]

biology and life sciences forum MDPI

Proceeding Paper

Development of Hydrogel-Type Jam with Chia (*Salvia hispanica* L.) Mucilage, Blueberry (*Vaccinium corymbosum*), and Cushuro (*Nostoc sphaericum*) †

Ignacio A. Albujar * and Stefano Málaga

Carrera de Ingeniería Industrial, Universidad de Lima, Av. Javier Prado Este 4600, Surco, Lima 15023, Peru; stefanomalaga3@gmail.com

* Correspondence: ignacio.albujar13@gmail.com

† Presented at the VI International Congress la ValSe-Food, Lima, Peru, 23–25 September 2024.

Abstract: In Peru, overweight and obesity affect 20–38% of adults, increasing the risk of NCDs (type 2 diabetes, heart diseases, and others) that emphasize the need for healthy foods. Chia (*Salvia hispanica* L.) seeds contain high amounts of polyunsaturated fatty acid essentials (omega-3) (17–23%), antioxidants, proteins, and minerals that prevent NCDs. Chia grows in the regions of Arequipa and Puno–Peru, with 4098 tn of production in 2023. Chia mucilage is a soluble fiber with a high water-holding capacity that possesses the techno-functional properties that would improve the properties of gelification and emulsification of foods: jams, ice cream, yogurt, and others. Peru holds the N°1 position in the ranking of blueberry (*Vaccinium corymbosum*) exporters. This berry contains antioxidants and flavonoids. Cushuro (*Nostoc sphaericum*) is a gelatinous spherical blue-green alga; it grows over 3000 masl on the Peruvian highland, and it contains good protein and polysaccharide contents. The work aimed to develop a hydrogel-type jam with chia mucilage (0.05–1.00%), blueberries (36–40%), and fresh cushuro (54–60%), compared with a control sample containing pectin and sugar. The characterization of the hydrogel-type jam was moisture (79.53 ± 1.51%), ash (0.20 ± 0.01%), protein (1.02 ± 0.28%), total carbohydrates (19.05 ± 1.76%), fat (0.21 ± 0.03%), antioxidants (318.56 ± 61.5 µm Trolox/g), and phenolic content (2.43 ± 0.93 mg GAE/g). Then, after 30 days of storage, the °Brix (9.9 ± 0.3), viscosity (3921.62 ± 1373.19), pH (3.18 ± 0.02), and water activity (0.82 ± 0.5) values of the hydrogel type jam complied with the Peruvian applicable legislation (NTP 203.047) and health law (No. 30021). The hydrogel's functional properties could help reduce the percentage of NCD, promoting the food industry with healthy products.

Keywords: chia mucilage; hydrogel-type jam; cushuro; blueberries; low calorie; functional

1. Introduction

In contemporary times, the increasing awareness of the importance of a balanced diet has heightened interest in functional foods, which not only fulfill basic nutritional requirements but also contribute to the enhancement of overall health and well-being. This trend has emerged as a response to the global prevalence of non-communicable diseases (NCDs), such as obesity, which is linked to the excessive intake of saturated fats, sugars, and sodium [1]. With the rising incidence of overweight, there is a growing demand for healthier food products, thus driving innovation in the development of functional foods. Chia (*Salvia hispanica* L.), grown in the Arequipa and Cusco regions, Peru, is known for itsrich nutritional profile in omega-3 fatty acids, dietary fiber, and protein [2]. Cushuro (*Nostoc sphaericum*) is a cyanobacterium that performs as a microalga that grows in the marine environments of the high Andean areas of Peru (>3000 masl). It has a high protein and iron content [3]. Blueberries (*Vaccinium corymbosum*) are grown in the Lima and Cajamarca regions, and they contain significant amounts of antioxidants, vitamins C and B, and phenolic compounds, which are vital for mitigating the risks of NCDs. The objective of this research was to develop and characterize a hydrogel-type jam with chia mucilage, antioxidants from blueberries, and fresh cushuro.

2. Materials and Methods

2.1. Raw Materials

Cushuro (*Nostoc sphaericum*) was harvested in the department of Ancash, province of Huaraz, district of Churup at the Churup lagoon. The chia (*Salvia hispanica* L.) and the blueberry (*Vaccinium corymbosum*) were obtained at the local market in Lima.

2.2. Extraction of Chia Mucilage

The extraction was performed according to the method described by Castañeda-Cachay et al. [4] with some modifications. The chia seeds were hydrated at 80 °C with agitation for 2 h in a stirring hotplate to extract the chia mucilage (CM). The mixture was dried in an infrared dryer (IRC D18, Inconfort, Seville, Spain) for 24 h at 50 °C at the Laboratorio de Alimentos Funcionales of the Universidad de Lima—Peru, ground with a mortar and pestle, and stored in aluminized bags at room temperature for further use.

2.3. Hydrogel Development

Four samples with different amounts of CM were developed and compared with the control sample (pectin and sugar) (Table 1). The cushuro and blueberries were weighed and separated into halves. The first half was liquefied and mixed with the second half (4 °Brix), then the mixture was cooked for 60 min at 90 °C (8 °Brix), the CM was added and cooked at 90 °C until the hydrogel-type jam (HJ) was obtained (11 °Brix). It was packaged, cooled, and stored in refrigeration (15 °C) until further analysis. The formulations complied with the Peruvian Technical Norm (NTP 203.047).

Table 1. Formulations for the development of hydrogel-type jam (HJ) with chia mucilage, blueberries and cushuro.

Samples	Cushuro (%)	Blueberry (%)	Sugar (%)	Pectin (%)	Sweetener (%)	Chia Mucilage (%)
Control	54.55	36.21	9.21	0.03	-	-
HJ 1	59.39	39.57	-	-	0.03	1.00
HJ 2	59.76	39.91	-	-	0.03	0.30
HJ 3	59.64	40.17	-	-	0.03	0.15
HJ 4	59.41	40.51	-	-	0.03	0.05

2.4. Optimization of Chia Mucilage Concentration

Chia mucilage (CM) was extracted using different chia:water ratios: CM1 (1:40), CM2 (1:30), and CM3 (1:20), with agitation at 200 RPM and 80 °C for 2 h according to Castañeda-Cachay et al. [4] with some modifications.

2.5. Viscosity

The viscosity property was evaluated 24 h after the samples were prepared according to Jeong et al. [5] with some modifications, using the Rheometer (MCR 92 Viscosimeter Anton Paar, Graz, Austria) with plate-plate geometry (50 mm diameter and 1mm gap).

2.6. Hydrogel-Type Jam Characterization

2.6.1. Proximal Composition

The proximal composition was determined according to official methods [6] with some modifications. The moisture content was determined at 110 °C to constant weight. The total protein content was determined as % nitrogen × 6.25 factor using a Kjeldahl analyzer (UDK 139, VELP, Usmate Velate, Italy). The ash content was determined by incineration at 600 °C for 14 h in a muffle furnace. The fat content was determined with hexane for 4 h. The measurements for the proximal assay were performed in triplicate.

2.6.2. Total Phenolic Content (TPC)

The total phenolic content (TPC) was determined by the Folin–Ciocalteau method [6], adding to each sample 4.5 mL of Methanol, 2.5 mL solution of 0.2N Folin–Ciocalteu, and 2.0 mL of 20% Sodium Carbonate. The absorbance of the samples was obtained at 760 nm using a spectrophotometer (UV 1280 Vis Spectrophotometer Shimadzu, Kyoto, Japan). The results were expressed as μg of Gallic Acid Equivalent (GAE)/g HJ. The analyses were performed in triplicate and presented as mean values ± SD.

2.6.3. Antioxidant Activity

The antioxidant activity was determined by the DPPH method [6] with some modifications at 517 nm by spectrometry (UV 1280 Vis Spectrophotometer Shimadzu, Kyoto, Japan). The results were expressed as μg Trolox/g HJ. The analyses were performed in triplicate and presented as mean values.

2.7. Sensory Analysis

The sensory evaluation was performed with 30 panelists using a 9-point hedonic scale from 1 (I dislike it very much) to 9 (I like it very much) with a mean value of 5 (I neither like it nor dislike it) according to Casquibol et al. [6].

2.8. Statistical Analysis

The results were expressed as mean ± standard deviation. All the measurements were determined in duplicate or triplicate. An analysis of variance (ANOVA), which was used to analyze data, was acquired at a 95% significance level with the Minitab 19.0 Software (Minitab Inc., State College, Palo Alto, CA, USA).

3. Results and Discussion

3.1. Optimization of Chia Mucilage Concentration

Chia mucilage (CM3, 1:20) showed the highest chia mucilage concentration, with an efficiency of 92.17%. The mucilage obtained was isolated and dried, and the moisture content was 8.27% and corresponded to 9.6% of the seed weight. Similar results were reported with *Linum usitatissimum* mucilage by Castañeda-Cachay et al. [4], at temperatures of 85 to 90 °C, with seed:water ratio(1:20),with an efficiency of 9.73%. Table 2 shows the results of the viscosity and sensory analysis of hydrogel samples. Sample HJ 3 was selected for its sensory characteristics which are very similar to the control sample, with an acceptability of 7.4 ± 0.96.

Table 2. Viscosity and Sensory Analysis of the Hydrogel-type jam (HJ) with chia mucilage, blueberries and cushuro.

Samples	Chía (%)	Viscosity [mPa·s]	Color	Smell	Texture	Flavor	General Acceptability
Control	-	657.13 ± 11.95 [c]	8.50 ± 0.52 [a]	8.50 ± 0.52 [a]	8.7 ± 0.48 [a]	8.5 ± 0.52 [a]	7.5 ± 1.08 [a]
HJ 1	1.00	563,578.5 ± 1497.94 [a]	8.1 ± 0.31 [ab]	8.2 ± 0.42 [a]	4.9 ± 0.56 [d]	5.2 ± 0.42 [c]	5.7 ± 0.94 [b]
HJ 2	0.30	512,377.05 ± 1514.44 [b]	7.8 ± 0.63 [b]	8.1 ± 0.87 [a]	3.9 ± 0.56 [e]	5.1 ± 0.73 [c]	6.3 ± 1.15 [ab]
HJ 3	0.15	369.49 ± 6.12 [c]	8.4 ± 0.51 [ab]	8.4 ± 0.51 [a]	8.0 ± 0.66 [b]	8.0 ± 0.66 [a]	7.4 ± 0.96 [a]
HJ 4	0.05	392.99 ± 6.98 [c]	8.0 ± 0.47 [ab]	7.9 ± 0.56 [a]	6.2 ± 0.41 [c]	6.9 ± 0.56 [b]	4.1 ± 0.73 [c]

The results are expressed as mean ± SD ($n = 10$). a, b, c, d, e values in the same column with different letters differ significantly when $p < 0.05$.

The viscosity of samples HJ 1 (563,578 ± 1497 mPa·s) and HJ 2 (512,377 ± 1514 mPa·s) were higher than control sample (657 ± 11 mPa·s) at a 95% level of confidence, and samples HJ 3 (369 ± 6 mPa·s) and HJ 4 (392 ± 6 mPa·s) were lower than control sample. HJ 3 scored (8.4 ± 0.5) for color, placing it in the same statistical group as the control., The acceptability of sample HJ 3 were statistical higher than other samples and it was evaluated with better sensory characteristics and similar to control sample.

3.2. Characterization

According to Table 3, moisture (79.53 ± 1.51%) content was higher than the control sample. The protein (1.02 ± 0.28%) content was significantly higher than the control sample (0.82 ± 0.12%), showing an increase of 19%, very similar to the reported by Nduko et al. [7]. Ashes (0.20 ± 0.01%) and fat (0.21 ± 0.03%) were higher than the control sample, with a low carbohydrate content (19.05 ± 1.76%) besides the contribution of chia [2] and cushuro [3] respectively.

Table 3. Proximal composition, antioxidant activity and total polyphenol content (TPC) of hydrogel-type jam (HJ 3).

Samples	Moisture [%]	Ashes [%]	Protein [%]	Fat [%]	Total Carbohydrates [%]	Antioxidants [μm trolox/g]	TPC [mg GAE/g]
Control	76.17 ± 1.11 ^{b}	0.14 ± 0.02 ^{b}	0.82 ± 0.12 ^{a}	0.10 ± 0.02 ^{b}	22.78 ± 1.03 ^{a}	265.02 ± 46.32	2.28 ± 0.74
HJ 3	79.53 ± 1.51 ^{a}	0.20 ± 0.01 ^{a}	1.02 ± 0.28 ^{a}	0.21 ± 0.03 ^{a}	19.05 ± 1.76 ^{b}	318.56 ± 61.50	2.43 ± 0.93

The results are expressed as mean ± SD (n = 3). a, b values in the same column with different letters differ significantly when $p < 0.05$.

The antioxidant activity of the sample HJ 3 (318.56 ± 61.50 μm Trolox/g HJ) was higher than the control sample (265.02 ± 46.32 μm Trolox/g Control) due to the high content of antioxidants present in the blueberries (6.1–36.6 μm Trolox/g) [8]. The total polyphenol (2.43 ± 0.93 mg GAE/g dm) content was statistically higher than the control sample (2.28 ± 0.74 mg GAE/g dm); according to Diaconeasa et al. [8]. Figure 1 shows the results of the storage of the formulation HJ 3 during 30 days. pH values ranged from 3.18 to 3.52 similar to the control sample (3.15 to 3.44) complying with the Peruvian Technical Norm [9] (NTP 203.047) (pH 3.0 to 3.8). However, the °Brix (7.6–9.9) in the HJ 3 sample was lower than the control sample (24.3–24.9), it is lower than the established by the NTP (65 °Brix) [10], due the fact that stevia was used as sweetener instead of sugar. Viscosity values (395.73–3921.62 UNIDADES) were higher than control sample (654.26–1979.64 UNIDADES), due to the properties of gelification and emulsification from chia mucilage (Atik et al.) [2]. The values of the water activity (0.80–0.82) was similar to the control sample (0.78–0.80) according the formulation developed.

The developed hydrogel (HJ 3) presented an intense purple color and a subtle sweet aroma characteristic of blueberries. For the sensory evaluation, 73% of the panelists rated its acceptability with a score above six (neither like nor dislike). The appearance, taste, texture, color, and aroma characteristics of the selected hydrogel sample (HJ 3) had average values of 6.47, 6.43, 7.00, 7.80, and 6.20, respectively, on the hedonic scale.

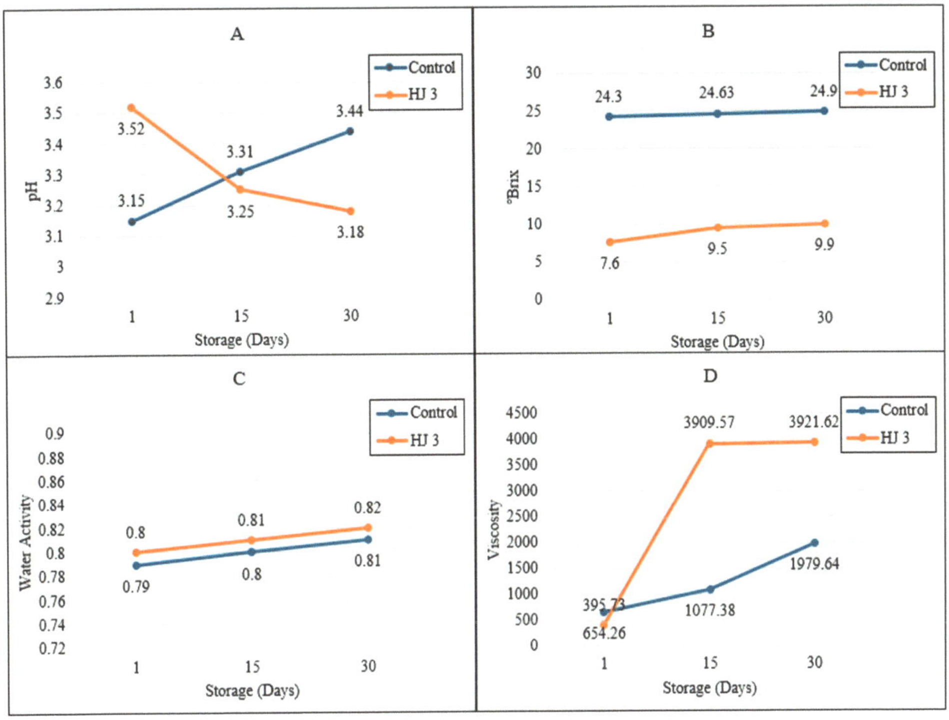

Figure 1. Shelf life of hydrogel-type jam (HJ 3) with chia mucilage, blueberries and cushuro. Effects of storage periods on pH (**A**), °Brix (**B**), water activity (**C**), and viscosity (**D**) from hydro-gel-type jam (HJ).

4. Conclusions

Chia mucilage contribuyed with consistency and viscosity to the hydrogel, similar to commercial jams. The gelatin-like hydrogel made with chia, stevia, cushuro, and blueberries presented high protein (1.02 ± 0.28%), low carbohydrate (19.05 ± 1.8%), high polyphenol(2.43 ± 0.93 mg GAE/g HJ) and antioxidants (318.56 ± 61.5 µm trolox/g HJ) content. HJ 3 is a healthy alternative because it contains lower caloric content due to its low °Brix and carbohydrate content, therefore can be an important complement to the family diet to promote healthy nutrition.

Author Contributions: Conceptualization, I.A.A. and S.M.; Methodology, I.A.A. and S.M.; Investigation and Data analysis, I.A.A. and S.M.; Writing—original draft preparation, I.A.A. and S.M.; writing—review and editing, I.A.A. and S.M. All authors have read and agreed to the published version of the manuscript.

Funding: This research received no external funding.

Institutional Review Board Statement: Not applicable.

Informed Consent Statement: Not applicable.

Data Availability Statement: Data is contained within the manuscript.

Acknowledgments: This work was developed at Laboratorio de Alimentos Funcionales de la Carrera de Ingenieria Industrial and supported by Universidad de Lima, Perú.

Conflicts of Interest: The authors declare no conflicts of interest.

References

1. Thapsuwan, S.; Phulkerd, S.; Chamratrithirong, A.; Gray, R.S.; Jindarattanaporn, N.; Loyfah, N.; Thongcharoenchupong, N.; Pattaravanich, U. Relationship between consumption of high fat, sugar or sodium (HFSS) food and obesity and non-communicable diseases. *BMJ Nutr. Prev. Health* **2024**, *13*, e000794. [CrossRef] [PubMed]
2. Atik, I.; Tekin Cakmak, Z.H.; Avcı, E.; Karasu, S. The effect of cold press chia seed oil by-products on the rheological, microstructural, thermal, and sensory properties of low-fat ice cream. *Foods* **2021**, *10*, 2302. [CrossRef] [PubMed]
3. Ponce, E. Nostoc: Un alimento diferente y su presencia en la precordillera de Arica. *Idesia* **2014**, *32*, 119–121. [CrossRef]
4. Castañeda-Cachay, A.P.; Zavaleta-Gutiérrez, N.E.; Siche, R. Optimización del proceso de extracción del mucílago de Linum usitatissimum utilizando un diseño secuencial. *Sci. Agropecu.* **2019**, *10*, 19–28.
5. Jeong, Y.; Lim, W.; Yoo, B. Relationship between Syringe Flow Measurements and Viscosity of Nectar-Thick Beverages for Dysphagia Management. *Foods* **2021**, *10*, 1981. [CrossRef] [PubMed]
6. Chasquibol, N.; Alarcón, R.; Gonzales, B.F.; Sotelo, A.; Landoni, L.; Gallardo, G.; García, B.; Pérez-Camino, M.C. Design of Functional Powdered Beverages Containing Co-Microcapsules of Sacha Inchi P. huayllabambana Oil and Antioxidant Extracts of Camu Camu and Mango Skins. *Antioxidants* **2022**, *11*, 1420. [CrossRef] [PubMed]
7. Nduko, J.M.; Maina, R.W.; Muchina, R.K.; Kibitok, S.K. Application of chia (*Salvia hispanica*) seeds as a functional component in the fortification of pineapple jam. *Food Sci. Nutr.* **2018**, *6*, 2344–2349. [CrossRef] [PubMed]
8. Diaconeasa, Z.; Iuhas, C.I.; Ayvaz, H.; Rugină, D.; Stanilă, A.; Dulf, F.; Bunea, A.; Socaci, S.A.; Socaciu, C.; Pintea, A. Phytochemical characterization of commercial processed blueberry, blackberry, blackcurrant, cranberry, and raspberry and their antioxidant activity. *Antioxidants* **2019**, *8*, 540. [CrossRef]
9. INACAL. *Norma Técnica Peruana NTP 203.047*, 1st ed.; Instituto Nacional de Calidad: Lima, Peru, 2017.
10. Codex Alimentarius Commission. Norma Para las Confituras, Jaleas y Mermeladas (CXS 296-2009). *Codex Alimentarius.* 2009. Available online: http://www.fao.org/fao-who-codexalimentarius (accessed on 15 March 2024).

MDPI AG
Grosspeteranlage 5
4052 Basel
Switzerland
Tel.: +41 61 683 77 34

Biology and Life Sciences Forum Editorial Office
E-mail: blsf@mdpi.com
www.mdpi.com/journal/blsf

www.ingramcontent.com/pod-product-compliance
Lightning Source LLC
Chambersburg PA
CBHW041344210926
51850CB00013B/2984
9783725841752